EurographicSeminars

Tutorials and Perspectives in Computer Graphics

Edited by W. T. Hewitt, R. Gnatz, and D. A. Duce

R. L. Grimsdale W. Straßer (Eds.)

Advances in Computer Graphics Hardware IV

With 124 Figures, 4 in Colour

Springer-Verlag
Berlin Heidelberg New York
London Paris Tokyo
Hong Kong Barcelona
Budapest

EurographicSeminars

Edited by W. T. Hewitt, R. Gnatz, and D. A. Duce
for EUROGRAPHICS –
The European Association for Computer Graphics
P. O. Box 16, CH-1288 Aire-la-Ville, Switzerland

Volume Editors

Richard L. Grimsdale
School of Engineering
University of Sussex
Brighton, BN1 9QT, U.K.

Wolfgang Straßer
Universität Tübingen
Wilhelm-Schickard-Institut für Informatik
Graphisch-Interaktive Systeme
Auf der Morgenstelle 10, C9
W-7400 Tübingen 1, FRG

ISBN-13: 978-3-642-76300-7 e-ISBN-13: 978-3-642-76298-7

DOI: 10.1007/978-3-642-76298-7

Softcover reprint of the hardcover 1st edition 1991

45/3140-543210 – Printed on acid-free paper

Preface

EUROGRAPHICS workshops on Graphics hardware have now become an established forum for an exchange of information concerning the latest developments in this field of growing importance. The first workshop took place during EG'86 in Lisbon. All participants in this event considered it a very rewarding workshop to be repeated at future EG conferences. This view was reinforced at the EG'87 Hardware Workshop in Amsterdam which firmly established the need for and a high interest in such a colloquium of technical discussion in this specialist area within the annual EG conference. The third EG Hardware Workshop took place in Nice in 1988 and this volume is a record of the fourth workshop at EG'89 in Hamburg.

The material in this book contains papers representing a comprehensive record of the contributions to the 1989 workshop.

The first part considers **Algorithms and Architectures** of graphics systems. These papers discuss the broader issues of system design, without necessarily raising issues concerning the details of the implementation.

The second part on **Systems** describes hardware solutions and realisations of machines dedicated to graphics processing. Many of these contributions make important references to algorithmic and architectural issues as well, but there is now a greater emphasis on realisation. Indeed many VLSI designs are described.

The contributions show that there is flourishing activity in the development of new algorithmic and architectural ideas. In particular, the interaction between the opportunities presented by VLSI technology, newly developed algorithms and the increasing diversity of applications encourage new solutions and keep graphics hardware a lively working area.

Finally, we should like to thank the members of the Programme Committee for their efforts invested in the planning of the workshop, the EUROGRAPHICS Association for supporting the event through scholarships, the Universität Hamburg and Werner Hansmann for hosting the event so excellently, and Kontron Elektronic for their generous sponsorship. Last, but not least, our thanks go to all the authors of this volume for the careful preparation of their contributions.

Richard Grimsdale, Wolfgang Straßer
Brighton and Tübingen, October 1990

Fourth Eurographics Workshop on Graphics Hardware

Workshop co–chairmen

Prof. Dr. W. Straßer, Institut für Informatik, Universität Tübingen
Prof. R.L. Grimsdale, School of Engineering, University of Sussex

Workshop Programme Committee

A.A.M. Kuijk, Centrum voor Wiskunde en Informatica, Amsterdam
R. Möller, Bergische Universität, Gesamthochschule Wuppertal
S. Molnar, University of North Carolina
R. Rossignac, IBM Research, Yorktown Heights
Ch. Shaw, University of Alberta
A. Kaufman, State University of New York at Stony Brook
D. de Jong, Philips Research Laboratory, Eindhoven
R.L. Grimsdale, School of Engineering, University of Sussex

Table of Contents

Part 1. Algorithms and Architectures

Part 2. Systems

Part 1

Algorithms and Architectures

Towards a Taxonomy for Display Processors

Bengt-Olaf Schneider

Universität Tübingen
Wilhelm-Schickard-Institut für Informatik
Graphisch-Interaktive Systeme
Auf der Morgenstelle 10 / C9
D-7400 Tübingen, FRG
e-mail: igbengt@dtupev5a.bitnet

Abstract

Image generation for raster displays proceeds in two main steps: geometry processing and pixel processing. The subsystem performing the pixel processing is called *display processor*.

In the paper a model for the display processor is developed that takes into account both function and timing properties. The model identifies scan conversion, hidden surface removal, shading and anti-aliasing as the key functions of the display processor. The timing model is expressed in an inequation being fundamental for all display processor architectures.

On the basis of that model a taxonomy is presented which classifies display processors according to four main criteria: function, partitioning, architecture and performance.

The taxonomy is applied to five real display processors: Pixel-planes, SLAM, PROOF, the Ray-Casting Machine and the Structured Frame Store System.

Investigation of existing display processor architectures on the basis of the developed taxonomy revealed a potential new architecture. This architecture partitions the image generation process in image space and employs a tree topology.

CR Categories and Subject Descriptors:
B.4.2[Input/Output and Data Communications]: Input/Output Devices — Image Display
C.1[Processor Architectures]
I.3.1[Computer Graphics]: Hardware Architecture — Raster display devices

General Terms: Algorithms, Design, Performance

Additional Keywords and Phrases:
Raster Graphics, Display Processors, Taxonomy, Classification, Model, Partitioning, Architecture

*The author is now with the IBM T.J. Watson Research Center, P.O. Box 704, Yorktown Heights, NY 10598, e-mail: bosch@ibm.com.

1 Introduction

During the last decade many computer graphics systems have emerged. A lot of them achieved their often impressing performance by using specially designed hardware. Especially with the move to raster graphics the quest for powerful hardware accelerators became more and more urgent. This is because the rendering of a million pixels and more requires a lot of computing power.

The different requirements for cost, performance and functionality gave rise to a great variety of computer graphics systems. All these systems try to optimise the compromise between the often contradicting design constraints. That optimisation is a rather difficult task because the design space, i.e. the number of possible alternatives, is huge. With the advent of VLSI technology the design space grew even larger. On the other hand VLSI technology introduced new constraints that required new tradeoffs. For instance, in contrast to conventional technology, in VLSI it is favourable to design systems with many identical and simple elements instead of a few, very complex parts. The paper takes these aspects into account.

We will concentrate on a special piece of hardware for computer graphics systems: the *display processor*. The display processor is performing the low-level tasks in the image generation process, namely pixel generation. We will present a classification scheme for these processors. We hope that such a taxonomy will help in the design of display processors. By providing means to classify different architectures and their properties it will be easier to compare and judge different designs. Furthermore, the taxonomy can serve as a starting point for the specification of a new display processor. It should be noted here that a taxonomy cannot (and is not intended to) provide figures of merit that identify one architecture to be *better* than another one. It can only supply criteria along which an evaluation could take place. In other words: the taxonomy is only a means to describe the available design space.

The paper starts with a short review of earlier attempts to classify display processors. Afterwards we present our own approach. First, we develop a model for the display processor that takes into account both, functionality and timing properties of the display processor. This model serves as an implicit definition of the term *display processor* and as a guideline for the taxonomy which is introduced in the next section. The employed criteria for the classification are presented and discussed. The classification scheme is then applied to some published architectures. The last part of the paper proposes a new display processor architecture that is the result of a search for "white spots" in the taxonomy.

2 Related Work

In [**Abram et al. 1986**] [**Fuchs 1988**] [**Kilgour 1985**] [**Dew et al. 1985**] overviews of existing graphics architectures are given. They only make the distinction whether a system partitions the image generation in image space or in object space. In [**Kilgour 1985**] this is called object serial or pixel serial respectively.

In [**Reghbati et al. 1988**] this classification is refined further. It is determined in how many sets the objects (pixels) are divided and what is the maximum cardinality of these sets.

3 Modelling the Display Processor

3.1 General Description

Producing computer generated images involves two main steps. The first one, *geometric processing*, prepares an image described on a high level of abstraction for the display on a physical output device. This involves the transformation and projection of a scene into a normalised coordinate system which can be easily mapped onto the output device. These coordinates are called normalised device coordinates (NDC). Preparing an image description for display also requires to break down the objects in the scene into simple geometric shapes, that can be easily handled by the low-level hardware. These shapes are referred to as atomic objects or primitives. The geometric processing also comprises calculations for the illumination of the scene, e.g. vertex colours or shadow polygons. Furthermore, in some implementations also a part of the hidden surface removal takes place in this step, e.g. assignment of priorities to objects in the scene and backface culling.

In the second step the scene description in NDC is processed further to be displayed on the output device. Since we restrict ourselves to raster devices this step is called *pixel generation*, because the final pixel colour is determined during this step. To this end, some or all of the following tasks have to be fulfilled:

Scan Conversion actually maps the primitives from NDC to PDC (Physical Device Coordinates), i.e. those pixels are identified which are covered by the considered primitives. This step is always part of the pixel generation process.

Hidden Surface Removal identifies those portions of a primitive that are visible to the viewer, i.e. that are not obscured by other objects. Usually, this problem is also solved during the pixel generation.

Shading. The colour of the different objects at the single pixels is a function of the optical properties of the objects and the spatial arrangement of objects and light sources in the scene. Shading computations approximate this complex relationship. The complexity of the shading calculations depends strongly on the actual implementation. Some systems perform only flat shading, whereas others implement computationally more complex shading models. There are several ways to partition the steps involved in the shading calculations between the geometric and the pixel processing.

Anti-aliasing. If no special care is taken raster images show artifacts that stem from the limited spatial sampling frequency. Anti-aliasing tries to alleviate these effects. Although it is commonly agreed that proper anti-aliasing is necessary, not all graphics systems are able to compensate aliasing effects.

We call that part of an image generation system performing the pixel generation the *display processor*. Different names for the display processor are pixel generator, display controller, or rasteriser. We prefer the term display processor because it reflects the fact that nowadays these pieces of hardware are fairly complex and also show a limited programmability.

It should be noted here that it is of course possible to implement most of the tasks for pixel generation in software. However, this would result in a system with low performance. We will not further investigate that alternative here.

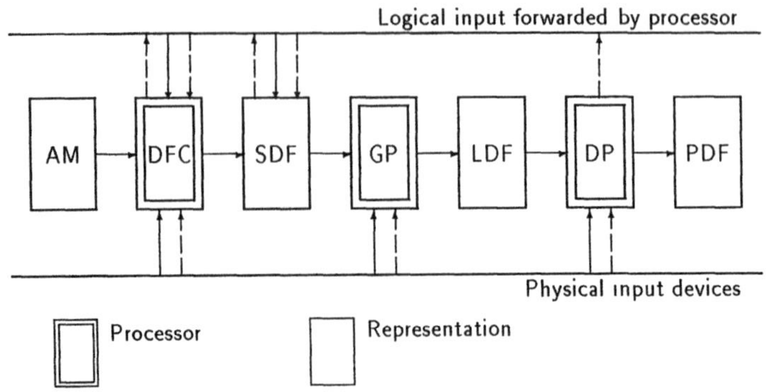

Figure 1: **Functional model of an image generation system.** AM – application model, SDF – structured display file, LDF – linear display file, PDF – physical display file, DFC – display file controller, GP – geometry processor, DP – display processor. The solid arrows symbolise input used to modify output. The dashed arrows denote input forwarded to other processors.

3.2 A Functional Model of the Display Processor

In order to get a clearer understanding of the tasks and functions of the display processor, we want to look closer at the underlying display processor model. We will try to give a somewhat formal description of its functions.

3.2.1 Output

Figure 1 shows a simplified functional model of an image generation system [**Carlbom 1980**]. A similar model has been proposed in [**Kilgour 1981**]. Both models partition the image generation process according to the properties and the levels of abstraction of the intermediate image descriptions.

Starting from the application model (AM) the scene is translated into a structured and often hierarchical description (SDF). This description is then transformed into a linear, non-hierarchical display file (LDF). The LDF is a representation of the scene that is based on the atomic graphical objects. They are described in the NDC system. Finally, the LDF is translated to the physical display file format (PDF).

As already pointed out in [**Carlbom 1980**] this model does not imply that there exist actual files that contain the LDF and PDF. The data contained in the LDF (PDF) description can either be transient. This means that they are generated and processed on the fly without intermediate storage. However, many graphics system have an explicit memory for the PDF: the frame buffer. Another possibility to implement the LDF or the PDF is to distribute them across several processing elements. This solution will be addressed later in the context of partitioning the display processor.

According to figure 1 the function of the display processor can be described by the following mapping:

$$LDF \overset{DP}{\longmapsto} PDF$$

The LDF is a low level description of the scene and consists of a sequence of object descriptions in the NDC system with associated shading information:

$$\begin{aligned} LDF \;::=\; & \{(p_i, s_i)\} \text{ where} \\ & type(p_i) \in P \\ & model(s_i) \in S \end{aligned}$$

p_i is the geometry description of the object i. The type of that object must be one out of the set P, that contains all primitives that can be handled by the display processor. Correspondingly, s_i is the shading information for object i. The shading model must be in the set S of all shading models that can be computed by the display processor.

The PDF is a pixel oriented description of the image. It can be modelled in the following way:

$$PDF ::= \{(x, y, C)\}$$

x and y denote the coordinates of pixels in the final image (PDC system). Without loss of generality we can restrict x and y to lie inside certain intervals: $x \in [0, X_{max}]$, $y \in [0, Y_{max}]$, where $X_{max} \times Y_{max}$ is the number of pixels on the screen. C represents the colour at the given pixel. The colour is usually represented in the RGB colour system. However, it may be possible to use other colour system, e.g. HLS or CMY [**Robertson 1988**].

Using these representations of the LDF and the PDF the function of the display processor is the following mapping:

$$(p_i, S_i) \stackrel{DP}{\longmapsto} (x, y, C)$$

This mapping together with the operations identified earlier lead to a conceptual algorithmic model for the display processor:

```
for (all objects and all pixels)
{ /* scan conversion */
  determine covered pixels ;
  /* hidden surface removal */
  determine visible pixels ;
  /* shading */
  determine colour of covered pixels ;
  /* anti-aliasing */
  perform filtering ;
}
```

The actual sequence of loops and operations is dependent on the implementation of the display processor.

3.2.2 Input

It is more difficult to model the behavior of a graphics system in reaction to external inputs than to model its output behavior. Although this is also true for display processors, the problem is less severe in this case. The only necessary input action to a display processor is to trigger a pick operation of an object at a certain pixel location. This means that the display processor must identify which primitive from the LDF is displayed at the specified pixel. We can therefore assume that the display processor accepts as an external input the inquiry for an object identifier at a certain pixel location. The reaction to this inquiry will be that after a certain delay the object identifier will appear at the display processor's output.

3.3 A Timing Model for the Display Processor

The functional model states that the display processor has to transform all primitives from the LDF to the PDF. This means that the display processor must

- execute *every* operation
- for *every* primitive in the LDF
- in order to give a colour to *every* pixel.

In order to meet the requirements set for the entire graphics system the display processor has to complete its traversal of the LDF in a certain time. Basically, the display processor has to produce pixels. Therefore, we define the time the display processor can spend for the generation of one pixel as a measure of its speed. We call this time the pixel write time t_{pw}.

As follows from the functional model this time depends on several parameters. First the length of the LDF is important, i.e. the number of primitives in the scene. We refer to this quantity as scene complexity C_{sc}. t_{pw} also depends on the properties of the primitives. We summarise these properties in the coefficient α_{sc} which comprises e.g. the average size of the primitives (relative to the total number of pixels on the output device) and their position relative to each other [1]. α_{sc} is strongly dependent on the employed scan conversion algorithm. If the actual algorithm is insensitive to the size of objects, α_{sc} is set to 1.

The display processor must handle all pixels. The more pixels there are the less time is available for each pixel. We call the number of pixels the image complexity C_I.

Obviously, the permitted pixel write time t_{pw} can be defined only if there is a limit for the image rendering time. This time, the frame time t_f, is an upper limit for the traversal of the LDF.

The pixel write time t_{pw} is then bounded by the following inequation:

$$t_{pw} \leq \frac{t_f}{\alpha_{sc} \cdot C_{sc} \cdot C_I} \tag{1}$$

Apart from the characteristics of the scene (C_{sc} and α_{sc}) and the image (C_I) the necessary speed of the display processor also depends on the kind and number of operations to be applied to the primitives. We call this parameter the functional complexity C_F. It represents the number of steps necessary to perform the operations. The maximum time available for each of these steps, t_{step}, is an indicator for the necessary clock frequency. It is given by the following inequation:

$$t_{step} \leq \frac{t_{pw}}{C_F} \leq \frac{t_f}{\alpha_{sc} \cdot C_{sc} \cdot C_F \cdot C_I} \tag{2}$$

We consider this to be the fundamental inequation for raster display processors. The actual values of its parameters give a good estimate of the performance of a particular display processor.

[1] The relative position influences how many objects are located on one pixel. This number is sometimes called depth complexity. The performance of some HSR algorithms or anti-aliasing algorithms depends on the depth complexity.

4 A Classification for Display Processors

Based on the model developed above we will now present a way to classify different display processors. We will extract as much information as possible from the model. However, some elements of the taxonomy cannot be deduced from the model. These topics, e.g. the topology of the architecture or its regularity, are empirical and stem directly from the investigation of existing display processors.

Moreover, we will point out some implications of the choices for one or another alternative.

4.1 A Functional Classification

The functional classification takes into account only the output behavior. The input behavior is not considered because it is of minor impact on the display processor architecture.

4.1.1 Primitive Types and Scan Conversion Algorithms

These two attributes of the display processor are closely coupled because the possible scan conversion algorithm has to comply with the available types of primitives. Therefore, one of the basic decisions in the design of a display processor is to define the set P of available primitive types. Subsequent to this decision suitable scan conversion algorithms can be chosen.

The choice of primitives also influences strongly the design of the geometry processor because it must break down the high-level description of the scene into primitives.

We will now look at some possible primitives.

Pixels are the simplest drawing primitives. Providing pixels as primitives offers the rest of the graphics system a well defined access to the output device. However, breaking down a scene into pixels is a computationally intensive task. (It is actually the scan conversion process.) Therefore this primitive is never the only primitive type provided.

Vectors are frequently used drawing primitives. This is for two reasons. First, historically the first graphics systems were pure vector drawers. This gave rise to a lot of graphics algorithms that worked best with vectors. Second, vectors are the most natural geometric shape for a lot of applications, e.g. technical drawings. Furthermore, vectors can often be generated faster than other drawing primitives.

Triangles are the simplest geometric shapes that cover an area on the screen. They exhibit some very pleasant properties. Triangles are always planar and convex. Linear interpolation of colours across triangles (Gouraud shading) is invariant with respect to rotations of the triangle.

Spans are trapezoids that are bounded by the upper and lower border of a scanline. The vertical edges can have an arbitrary angle. Usually, spans are supposed to be planar. Spans are often used to render general polygons by assembling these polygons from several spans. (Spans are also called scanline segments.)

General polygons form a generalisation of the triangle primitive. In contrast to triangles, general polygons can be concave, non-planar and self-intersecting. The result of Gouraud shading general polygons depends on the orientation of the polygon. Therefore it is advisable to use other shading algorithms with general polygons, e.g. biquadradic interpolation.

Sometimes polygons are restricted to be convex, planar and not self-intersecting. This simplifies considerably the scan conversion but it does not cure the shading problems.

Freeform surfaces are a very powerful means for object description. They offer a high degree of flexibility for controlling the shape of the surface. There are several kinds of surface representations around, interpolating and approximating ones. The more popular ones are Bézier surfaces, B-spline surfaces and NURBS surfaces.

The advantage of providing freeform surfaces as a display processor primitive is that the transformation of the high-level description of the scene into the LDF is rather simple. Furthermore, the LDF can be kept small compared to a triangle-based LDF.

On the other hand, scan conversion of freeform surfaces is difficult. Although there exist some proposals how to implement such a scan converter in hardware [**Pulleyblank et al. 1987**] [**Schneider 1988b**] this is not yet state of the art.

Halfspaces are the basic building blocks of constructive solid geometry (CSG). The boolean combination of linear and/or quadratic halfspaces can be used to generate many objects in a convenient way. The feasibility of display processors for halfspaces has been already demonstrated.

Usually, display processors can handle more than one primitive type. It is up to the designer to select those primitives that are most appropriate for the intended applications. It should be stressed again that the choice of the set of primitives is crucial because it has a strong impact on other parts of the display processor and the rest of the graphics system.

The choice of the scan conversion algorithm is a consequence of the selection of primitives. Scan conversion algorithms can be coarsely divided into two classes:

Contour tracking algorithms first identify the pixels on the borders of the primitive. Afterwards they fill all pixels in between the border pixels with the appropriate pixel values. Typical representatives for this class of algorithms are scan line algorithms that use DDA or Bresenham for computing the edges of triangles or polygons.

This method of scan conversion is applicable only to vectors, triangles, trapezoids (spans) and general polygons.

Inside testing algorithms examine (a subset of) all pixels whether they are inside the primitive under consideration. These algorithms are often very elegant from an algorithmic point of view. Unfortunately, they tend to test many pixels in vain because these pixels are outside of the primitive. For some primitives there exist algorithms that reduce the tests of empty pixels on the expenses of more complicated control structures in the algorithm [**Pineda 1988**].

Algorithms that employ inside tests are available for all primitives. They are mandatory for scan conversion and hidden surface removal using ray tracing.

One important difference between these two classes of scan conversion methods is in which order they generate pixels. The sequence of pixels produced by contour tracking algorithms is defined by the geometry of the object to be displayed. It is therefore an unpredictable sequence of coordinates which requires the display processor to be granted random access to the output device. In particular, this means that there has to be a frame buffer.

In contrast, for inside testing algorithms the sequence of pixels to be generated can be defined. It is therefore possible to produce the pixels in the scan order of the output device. Hence, pixels may be written directly to the output device without any intermediate buffering. That becomes important when the display processor is integrated into a real-time graphics system.

4.1.2 Hidden Surface Removal (HSR)

The decision about which parts of an object are visible to the viewer is an essential step for the creation of realistic images. Plenty of algorithms have been presented to solve the visibility problem, for both vector and raster displays [**Sutherland et al. 1974**] [**Latham 1985**] [**Lakshminarasimhan et al. 1989**].

The choice of the HSR algorithm determines which capabilities the display processor will have. Not all HSR algorithms are equally well suited to handle e.g. penetrating objects or transparency. Moreover, the memory requirements vary significantly for the different algorithms.

As pointed out in [**Sutherland et al. 1974**] HSR is essentially a sorting process. Therefore, many HSR algorithms require an extensive sorting step prior to the actual HSR. The order of sorting in the directions of x, y and z is unimportant. However, the various algorithms may be of different efficiency with regard to the amount of coherence exploited during the sorting.

A major distinction between different HSR algorithms is whether they use priorities or real depth for the visibility decision. The priority is assigned to an object once. In contrast, the depth is changing across the object and has to be computed for every pixel.

Since priorities define a partial order over the objects it is impossible to handle cases of penetrating objects. Such constellations must be eliminated by subdividing the objects along the intersections.

According to [**Lakshminarasimhan et al. 1989**] HSR algorithms fall into three major categories.

Object space algorithms. [2] HSR in object space is usually done in the geometry processor. The result is a number of primitives that are non-overlapping in object space. They are forwarded to the display processor. Hence, the display processor is not concerned with the HSR.

List-priority based algorithms in fact solve the visibility problem outside the display processor. The primitives are presorted according to their priority (either in priority order or in reverse priority order). The display processor then writes the primitives in that order to the output device. Hence, the HSR in the display processor is quite easy.

The *painter's algorithm* is based on a reverse priority ordering of the objects. This guarantees that the object with highest priority is written last and therefore lies in

[2] In the context of HSR *object space* means that the accuracy of the visibility calculations depends on the precision of the object representation in the SDF. This meaning of *object space* must not be confused with the notion of *object space partitioning* defined in this paper.

front. Translucency can be incorporated into the painter's algorithm easily by mixing the object's colour with the previous pixel colour. Anti-aliasing is not straightforward; problems exist to handle the inner edges of polygon meshes.

When using a *sorting in priority order* the frontmost object is displayed first. The pixels covered by that object are tagged. Objects with a lower priority can only write to untagged pixels. Thus, each pixel is written only once. This reduces the number of accesses to the display significantly compared to the painter's algorithm. Both, anti-aliasing and translucency are non-trivial tasks for this method.

Image space algorithms. [3] solve the visibility problem with a precision that depends on the screen resolution. According to [**Lakshminarasimhan et al. 1989**] there are four basic types of such HSR algorithms:

- **Scanline algorithms** have been developed mainly to handle polygons together with contour tracking scan conversion algorithms. They first identify for every scanline which objects are contributing to this scanline. The edges of these objects are kept in an active-edge-table. For every scanline, these edges are sorted in x (scan order). Every time an edge is crossed while scanning a scanline it is decided which of the active polygons is nearest to the viewer.

  If the algorithm is implemented as sketched out, it is unable to process penetrating objects; it is then mainly priority based. However, it is possible to extend the algorithm to handle such cases. Translucency can be incorporated in scanline algorithms.

- **Depth-buffer** is the most popular HSR technique in today's high-performance graphics systems. For every pixel, depth and colour are stored. Every object writes its colour only to those pixel whose depth is greater than the object's depth. Thus, the visibility problem is solved pixel by pixel. Hence, the depth buffer algorithm is able to handle intersecting objects without extra effort.

  Transparent objects can be treated correctly only if they are written to the buffer last. This requires an extra sorting step. It is very difficult to incorporate good anti-aliasing into the depth buffer-algorithm. However, there exist extensions to the depth-buffer that address this problem, e.g. [**Carpenter 1984**] [**Porter et al. 1984**].

- **Area sorting** subdivides the screen until a specified level of subdivision has been reached or the whole subscreen is covered by only one primitive. The quadtree representation of images uses this algorithm.

  Incorporation of translucency into this concept is a non-trivial task. There are difficulties to produce anti-aliased pictures with the area sorting approach.

- **(Primary) ray tracing** is a technique which shoots through every pixel a ray of light into the scene. The object hit first by the ray is visible to the viewer. This technique requires a scan conversion algorithm of the inside test class. Note that ray tracing has been developed to model effects like multiple reflections and refractions of light at objects. The computational complexity only pays off if this is desired. Ray tracing can easily handle transparency and penetrating objects. Also anti-aliasing can be integrated readily into this method.

---

[3] The term *image space* used in the context of HSR differs from that introduced later on in this paper. Here *image space* only means that the precision of the HSR is a consequence of the output device.

4.1.3 Shading

In order to give computer generated images a realistic appearance the illumination of the scene must be modelled. Since the circumstances of illumination in the real world are very complex shading of scenes in computer graphics always simplifies the physical world to some extent.

In general, shading in computer graphics means to compute the colour of an object at certain points. The method, that computes this colour is called *shading model.* The *shading algorithm* specifies at which points of the object the shading model is applied. (The distinction between models and algorithms is somewhat arbitrary. Models and algorithm are often related to one another.) Since the very beginning of computer graphics shading models and algorithms have been proposed. A good survey of models and algorithms is given in [**Claussen 1988**]. Here, we will not discuss the various shading models. Only the most important shading algorithms will be enumerated.

Wash shading is the simplest algorithm. It assigns a fixed colour to every primitive. This colour does *not* depend on any light sources in the scene. This algorithm is very simple but does not give realistic images.

Flat shading assigns one colour to the whole primitive. In contrast to the wash shading algorithm this colour has been calculated in dependence of the illumination of the scene.

Gouraud shading interpolates colours from one point of the primitive to another point. The colours at the two points are computed in the geometry processing step.

Usually Gouraud shading employs linear or bilinear interpolation of colours from one vertex of a polygonal primitive to another vertex. The shading model used to compute the vertex colours can be chosen freely.

Phong shading specifies that the normal vector of a primitive has to be interpolated from one point on the primitive to another point. For every step the interpolated normal vector together with other attributes of the object (reflectance, glossiness, etc.) is used to compute the object's colour.

As with Gouraud shading, Phong shading was originally intended for polygonal primitives. Then, the normal vector is (bi)linearly interpolated from vertex to vertex. Although, there is a Phong shading model this model is not necessarily tied to the Phong algorithm.

Ray tracing is (apart from determining visible objects) used to find out where shading has to take place. Since ray tracing can provide much information about the scene's illumination very demanding shading models can be employed.

4.1.4 Anti-aliasing

Aliasing is a consequence of determining the colour of the entire pixel by sampling the pixel only at its center or too few sampling points in the pixel. In static pictures these anomalies are the well known jagged edges and thin lines broken into parts. More deficiencies appear in animated pictures: small blinking objects, objects jumping from scanline to scanline and "running ants" on jagged edges.

Although, in practice, a perfect anti-aliasing is impossible [**Blinn 1989a**] [**Blinn 1989b**] there are ways to cure the worst aliasing effects. This can be done either by increasing the sampling frequency or by taking into account that a pixel covers a non-zero area. The latter methods try to compute how much of a pixel is covered by the contributing objects.

The various methods differ in the number of passes they take to produce the anti-aliased image. There are one-pass and two-pass techniques. Two-pass techniques first render the entire image and generate additional information about the image, e.g. contribution of an object to a pixel. In the second pass this information is used to enhance the picture. In contrast, one-pass techniques try to approximate this additional information. They generate the final pixel colour immediately. One-pass techniques exhibit serious drawbacks if the LDF is not preprocessed properly, e.g. sorting the objects from front to back.

A further distinction between different anti-aliasing techniques is whether the anti-aliasing is adaptive or not. Anti-aliasing is necessary only at the contours (edges) of objects. However, many implementations do not explicitly search for edges but process the whole picture in a uniform manner, e.g. supersampling.

Processing only the contours of objects is correct only if there are no intersections of objects. Some anti-aliasing techniques are not able to handle penetrating objects.

We will now describe some common anti-aliasing techniques in display processors.

The gz-buffer is a one-pass technique [**Ghazanfarpour et al. 1987**]. It is basically a depth buffer with another parallel geometry-buffer, the g-buffer. The g-buffer contains information about the current coverage of a pixel. This information is used if the edge of another object is running through that pixel. Then, the colour of one of the four neighbor pixels is mixed with the colour of the object in order to obtain the new pixel colour. Also the new contents of the g-buffer is computed.

The gz-buffer is able to handle penetrating objects. It is restricted to at most one edge, i.e. two objects, per pixel.

Compositing is a one-pass technique, too [**Porter et al. 1984**]. For every pixel in addition to the colour an alpha channel is sustained. The value of alpha represents the fraction of the pixel that is covered by a particular object. Also the depth of the object at the four corners of the (square shaped) pixel is evaluated.

Correspondingly, for every pixel the depth values at its four corners are stored. Using the depth values at the pixel corners it is determined how the new and the old pixel overlap each other. The value of alpha and the pattern of overlap are combined to the new pixel colour and the new alpha value.

Supersampling means that one pixel is sampled at several points. For each of these subpixels depth and colour are computed. That high resolution picture is then compressed to the resolution of the output device by averaging all subpixels contributing to a pixel. Although supersampling is still a point sampling technique it produces very good results since the aliasing effects are diminished by the averaging step. Supersampling can be easily combined with all scan conversion, shading and HSR algorithms because the averaging step is completely independent from the image generation step. Unfortunately, supersampling is very expensive with respect to both time and memory.

The A-buffer is an extension of the depth buffer [**Carpenter 1984**]. Instead of storing only depth and colour of the frontmost object the A-buffer stores depth and colour of all objects that are potentially contributing to a particular pixel. (The depth of the objects has been determined at the pixel center.) These objects are stored in a depth sorted list. For every object in the list a subpixel mask is computed. The subpixel mask specifies which subpixels are covered by an object.

In a second step, for every subpixel the visible object and its colour are computed. These subpixel colours are then averaged in order to obtain the final pixel colour.

The A-buffer technique has the drawback that its requirements for time and memory are not constant. They depend on the statistical properties of the scene to be rendered.

Filtering is not a real anti-aliasing technique. It combines certain pixels in the neighborhood of the pixel under consideration to determine the pixel colour. The effect of a low-pass filter is only a blurring of the picture.

4.2 Classification by Partitioning

The fundamental inequation for raster display processors (2) indicates that display processors have to parallelise the image generation process in order to achieve reasonable performance.

A short example will explain this: For a screen with 1000 × 1000 pixels, a scene with 10,000 primitives covering 1000 pixels each ($\alpha_{sc} = 0.001$) and a frame time of 0.1 s we obtain ($C_F = 1$):

$$t_{step} \leq 10ns$$

This time is just within the limits of today's technology. More complex scenes, screens with a higher resolution or shorter frame times will demand shorter step times. Therefore it is natural to search for ways to introduce parallelism into the display processor operation.

The number of steps necessary to generate an image is fixed. A partitioning describes how these steps can be distributed among a number of processor elements that are working in parallel on different steps. The result of a partitioning will be a rule how to construct such processor elements. The display processor is then constructed from *display sub-processors*. t_{step} for each of these sub-processors is greater than the step time for the display processor itself.

We will now discuss the different possibilities to partition the display processor. The inequation (2) indicates how the necessary partitioning can be accomplished. Every parameter stands for another domain of partitioning.

4.2.1 Partitioning in the Time Domain

In the inequation (2) the time domain is represented by the frame time t_f. Enlarging t_f for the display sub-processors means that each of these sub-processors has more time available to perform its operations. In order to sustain the performance of the display processor several sub-processors have to work in parallel on different, subsequent tasks. While one sub-processor is working on the task T_i the next sub-processor is processing task T_{i+1}.

4.2.2 Partitioning in Image Space

The image complexity C_I stands for the number of pixels. The time t_{step} can be enlarged for the sub-processors by distributing subsets of all pixels to the sub-processors, i.e. C_t is reduced.

There are several ways to achieve this. A single display sub-processor can either work on single pixels, on complete scanlines or on contiguous parts of the screen like stripes or windows. Another possibility to partition the screen is to interlace the pixel sets that the different sub-processor are treating [**Parke 1980**] [**Hu et al. 1985**]. The association of sub-processors and certain pixels can either be static or dynamic.

This partitioning concept implies that either all objects are broadcast to all sub-processors or that the objects are sorted according to that part of the screen they are contributing to.

The image space partitioning approach has potential drawbacks for images where the objects are not distributed evenly across the image. Then the workload may be unbalanced among the sub-processors.

Image space partitioning is often reflected in the organisation of the image memory, i.e. the frame buffer.

4.2.3 Partitioning in Object Space

Reducing the scene complexity C_{sc} means to off-load the sub-processors from a part of the objects. Every sub-processor, a so-called object processor, is handling a subset of the LDF. That subset can contain either a single primitive or several primitives. These primitives may be independent of each other. Alternatively, the primitives may all belong to a higher-order object that has been defined by the application (AM).

A potential problem with object space partitioning is that the processing time for different primitives may vary. (For instance, processing bigger primitives could take more time than small objects.) Since, in practice, these differences can be only estimated, some object processors may run idle while others are overloaded.

A special kind of object space partitioning is the partitioning in the parameter space. This strategy is frequently used when parametric surfaces must be displayed by a display processor that can only handle e.g. triangles.

4.2.4 Partitioning in the Functional Domain

The individual display sub-processors are less burdened by their task if they have to perform less operations in a certain time period. This means that the parameter C_F is made smaller. (C_F is the functional complexity and represents the number of steps to be performed for each pixel). This can be achieved by distributing the various operations amongst the display sub-processors.

The different sub-processors are not performing the same function and are therefore probably not of the same type [4]. This can be a drawback because the effort to design such a display processor is bigger than for the other partitioning strategies. The fact that there are different types of sub-processors leads to a lower regularity of the display processor (see below).

[4]It is possible to implement different functions on identical processor elements. This can be achieved e.g. by means of (micro)programming. A good example for such a solution is the system described in [**Clark 1982**].

4.2.5 Multi-level Partitioning

The partitioning schemes presented so far are pure solutions in the sense that they partition the display processor's task only in a single domain. In contrast, real systems often employ a combination of different problem partitioning approaches. We can distinguish here such solutions which divide the problem several times in the same domain and solutions that partition in different domains.

Looking at systems that employ a multi-level partitioning leads to the following observation: Each system partitions the image generation either in image space or in object space. We call these partitioning schemes *primary partitioning*. Partitioning the time domain or the functional domain are auxiliary or *secondary partitionings*. The secondary partitionings are used to efficiently implement one of the primary problem subdivision strategies.

4.2.6 Notation

We will now introduce a notation that expresses how a certain display processor partitions its task. This notation is an extension of the scheme proposed in [**Reghbati et al. 1988**].

We represent the domain in which the partitioning takes place by the first letter of its name (F, I, O, T). In parenthesis we put two arguments: The first tells in how many parts the domain is divided, i.e. how many sub-processors are working in parallel in that domain. The second parameter states the maximum number of elements in each partition, i.e. how many elements each sub-processor handles. Elements can be e.g. pixels, scanlines, or primitives.

For instance, the expression

$$I(4096, 256)$$

means that the display processor partitions the image space in 4096 subimages each containing 256 e.g. pixels.

Multi-level partitioning schemes are represented by catenating the different partitioning levels with a '$\circ$', e.g.

$$I(16, 64) \circ I(4, 64 \times 256)$$

can be the description of a system that divides the screen into 16 groups of 64 scanlines each. Each group is further partitioned into 4 parts that are handling 256 pixels of each of the group's 64 scanlines.

If there are different kinds of domain partitioning on the same level they are separated by a '$\|$'.

4.3 Classification by Architecture

A major issue for classifying display processor (and computers in general) is how the algorithms have been mapped onto a hardware structure. To categorise general computer architectures is a difficult task on its own. Several proposal have been made in the past [**Giloi 1983**]. However, they do not concentrate on what we consider to be important and descriptive for display processor architectures. We will therefore highlight only some useful features that are also generally accepted as descriptive for computer architectures. (All features assume that display processor is a multiprocessor.)

Topology of the network. The topology gives an idea about the physical construction of the processor. Possible topologies are pipelines, two-dimensional arrays, arrays in more than two dimensions (hypercubes), tree structures, etc.

Different architectures can be further distinguished by their capability to be configured dynamically, i.e.the topology can be adapted to different applications.

Control of the network. A multiprocessor network can be controlled by a central master. Alternatively, a decentralised control is distributed across a network of processors having equal rights. Furthermore, the control can be classified according to the manner how commands are distributed inside the network. The commands can be either broadcast simultaneously to all processor or propagated from one processor to the next.

The communication scheme is a property that is strongly related to the control structure. In the context of a VLSI design it is preferable to employ local communication structures instead of global ones.

Data types. Architectures can be distinguished with respect to the type of data propagated through the network. The classification of the data types falls into two categories. The first category is the kind of data, e.g. pixels, objects, colours, etc. The second category states how the data is represented. Possible representations are integers, floats, lists etc.

Type of parallelism. In a multiprocessor there can be parallelism in the execution of different commands and/or the processing of different data. Such processors are called single/multiple instruction machines with single/multiple data processing (SISD, MISD, SIMD, MIMD).

Parallelism can be further classified as coarse grain parallelism or fine grain parallelism.

Regularity. By regularity we understand the number of identical elements in relation to all elements. The more regular an architecture is the better it is suited for a VLSI implementation. We will now give a somewhat intuitive definition of regularity:

Let be N the number of all processing elements in the architecture and T the number of types of processing elements, e.g. the number of different chips. By n_i we denote the number of processing elements of type i.

We define the regularity R as the product of two factors r_1 and r_2:

$$
\begin{aligned}
r_1 &= 1 - \frac{T-1}{N} \\
r_2 &= \begin{cases} 1 - \sqrt{\frac{1}{T-1}\sum_{i=1}^{T}(\frac{1}{T} - \frac{n_i}{N})^2} & \text{für } T > 1 \\ 1 & \text{für } T = 1 \end{cases} \\
R &= r_1 \cdot r_2
\end{aligned}
$$

r_1 takes into account only the size of the architecture and the number of different chips. It exhibits the following properties.

- $T = 1 \Rightarrow r_1 = 1$
  The architecture is most regular if there is only one type of element.
- $N = T \Rightarrow \lim_{T\to\infty} r_1 = 0$
  If all elements are different the regularity is the smaller the more elements there are.
- $0 < r_1 \leq 1$

- The more types there are in the architecture the lower the regularity.
- The regularity is higher if there are more elements in the architecture.

r_2 reflects how many instances of each chip type there are in the architecture. We chose r_2 as the inverted standard deviation of the normalised population (n_i/N) of the type classes i. Hence it has the following properties:

- $n_i = n_j = \frac{N}{T} \; \forall \, 1 \leq i, j \leq T, \; i \neq j$
  $\Rightarrow \max(r_2) \;, \; r_2 = 1$
  If the number of elements in the different classes is the same, the regularity has a maximum.
- $n_1 = N-T+1, n_i = 1 \; \forall \, 1 < i \leq T$
  $\Rightarrow \min(r_2)$
  If the elements of the different types are distributed totally unequally, r_2 has a local minimum.
- $0 < r_2 \leq 1$

The first two properties of r_2 reflect the idea that structures look more regular if all elements appear with the same quantity.

4.4 Classification by Performance

Performance can be classified along different guidelines.

Characteristic parameters. The fundamental inequation (2) for raster display processors gives the maximum time for each operation of the display processor (t_{step}). This time can serve as a common basis for comparing and classifying the performance of different display processors. The bigger the time t_{step} may be (for equal functions) the easier such a display processor can be implemented with available technology.

Primitives per time period. It is very frequent to specify how many primitives can be processed by a particular display processor in certain period of time. Usually, this time period is either 1 second or one video frame. The video frame time is often set to 1/30 second or 1/60 second.

Delay time. The time between issuing a command to the display processor and obtaining visible results is called delay time. This time, although often not specified, is important to judge the ability of a graphics system to conveniently interact with the user.

5 Classification of some Existing Display Processors

This section applies the developed taxonomy to some display processors that have been published in the last years. The intention of this section is not to give a survey of existing display processors.

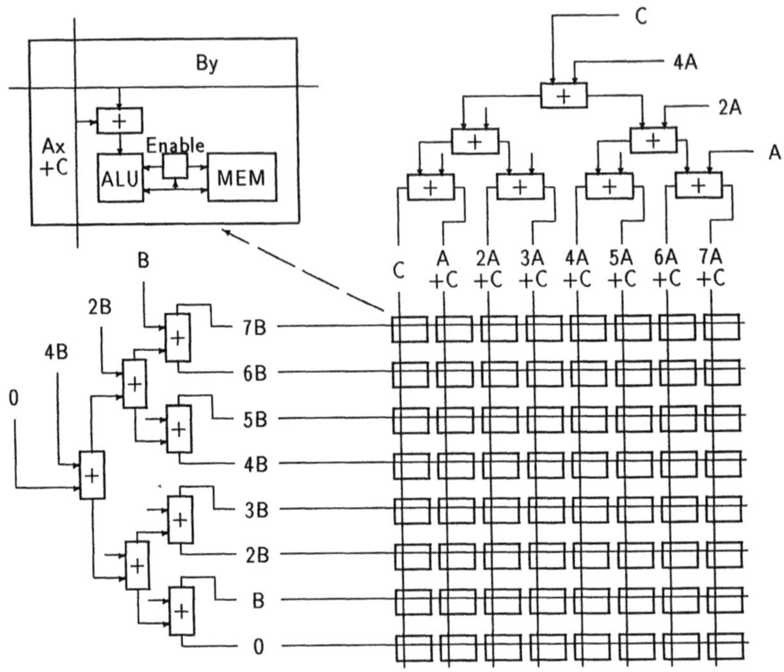

Figure 2: **Conceptual block diagram of the Pixel-planes display processor.**

5.1 Pixel-planes

The Pixel-planes system is one of the best documented display processors in the literature [**Fuchs et al. 1982**] [**Poulton et al. 1985a**].

Figure 2 shows the principal construction of the Pixel-planes architecture ([**Fuchs et al. 1985**]). It is essentially a frame buffer whose pixel cells have been enhanced by some extra logic. This logic enables Pixel-planes to compute the expression $Ax + By + C$ in parallel for all pixels (x, y). Evaluation of this linear expression with different sets of parameters A, B and C enables the Pixel-planes system to render z-buffered Gouraud-shaded polygons.

5.1.1 Functional Classification

Primitive Type(s). Convex, planar polygons [5].

Scan Conversion. Inside test based on a polygon representation by the bounding lines of the polygon.

[5] In [**Fuchs et al. 1985**] it is shown how the Pixel-planes system can be used to render circles and spheres.

Hidden Surface Removal Depth-buffer.

Shading. Gouraud shading (interpolation of *precomputed* vertex colours).

Anti-aliasing. None [6].

5.1.2 Partitioning

Pixel-planes parallelises the image generation process by one-level image space partitioning: $I(512 \times 512, 1)$.

5.1.3 Architecture

Topology. The two main building blocks of the Pixel-planes system are two binary trees of bit-serial adders and an array of pixel processors. The outputs (leaves) of the binary trees provide partial sums of the expression $Ax + By + C$ to the pixel cells.

Control. The whole Pixel-planes system is controlled by a central control unit on each Pixel-planes chip [**Poulton et al. 1985b**]. The controllers receive the commands from the outside world – either from the geometry processor that issues drawing commands or from the scan-out circuitry that reads out the image from Pixel-planes.

Data types. The data fed into Pixel-planes are always the parameters A, B and C of a linear expression, describing either edges, or depth or colours. A, B, C are integers and are supplied in bit-serial format to the Pixel-planes system.

Parallelism. The commands and parameters to Pixel-planes are the same for all pixel processors. The values of the linear expressions vary from pixel to pixel. Thus, all pixel processors are operating with different data. Therefore, Pixel-planes is a SIMD display processor.

Regularity. The Pixel-planes employs only one type of chip. It is therefore highly regular ($R = 1$). The Pixel-planes IV system is built from 2048 identical chips each of them storing 128 pixels [**Fuchs et al. 1988**].

5.1.4 Performance

Characteristic parameters. The following numbers are based on the assumption that Pixel-planes is rendering triangles. Then seven steps are required to render one triangle (3 edges + 1 depth + 3 colours). Some of the numbers are taken from the literature. It should be noted that the speed of Pixel-planes is independent of the actual size and location of the triangles ($\alpha_{sc} = 1$).

$$\begin{aligned} t_f &= 1\ s \\ \alpha_{sc} &= 1 \\ C_{sc} &= 35,000 \\ C_F &= 7 \\ C_I &= 1 \end{aligned}$$

Using the inequation (2) we obtain $t_{step} \leq 4\ \mu s$

[6] [**Fuchs et al. 1985**] explains how Pixel-planes can perform anti-aliasing by either supersampling or employing subpixel masks.

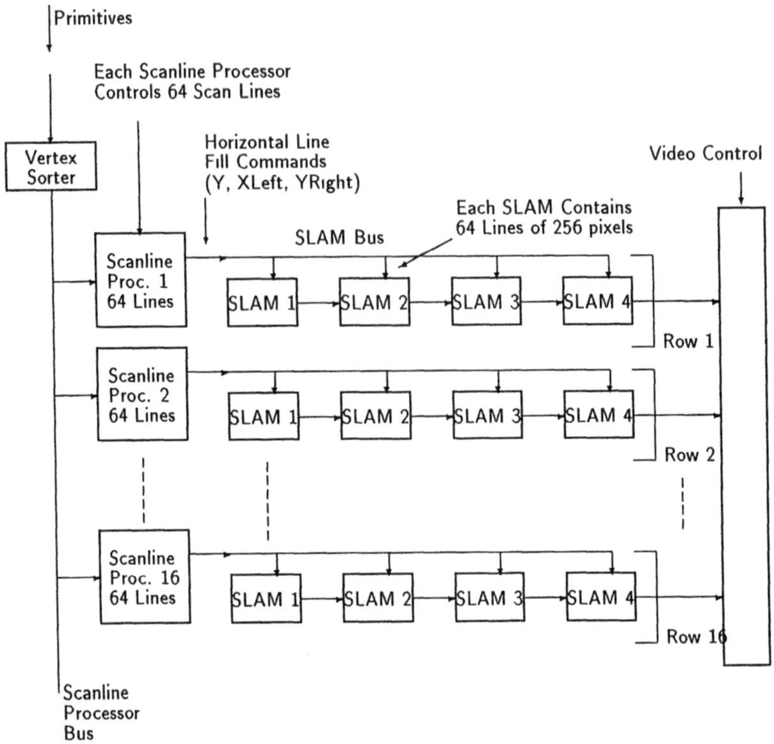

Figure 3: **Block diagram of the SLAM display processor.**

Primitives per second. Pixel-planes IV can process about 35,000 Gouraud-shaded, z-buffered triangles per second. Assuming that there are 30 frames per second, these are nearly 1,200 triangles per frame.

Delay time. The delay time of Pixel-planes is not specified in the literature. However, the working principle of Pixel-planes suggests that the delay is exactly the time necessary to process one polygon ($\approx 30\ \mu s$).

5.2 SLAM

In [**Demetrescu 1985**] a display processor has been presented that uses two different kinds of subsystems (figure 3): *Scanline processors* break down the incoming polygon descriptions into spans.

The spans are described by the scanline they are covering and the coordinates of the first and the last pixel in that scanline. Each scanline processor is responsible for 64 scanlines.

These spans are dispatched to the *scanline access memories* (SLAMs). Each SLAM fills the specified span with a downloaded pixel pattern. By specifying pixel patterns characters can be displayed. Each SLAM chip contains the memory of 64 scanlines with 256 pixels each. All 256 pixels in one scanline can be accessed, processed and stored in one step. SLAMs can be cascaded to construct longer scanlines.

5.2.1 Functional Classification

Primitive Type(s). Y-monotone polygons [7] and characters. (Lines are drawn as thin polygons.)

Scan Conversion. Contour tracking. The scanline processor computes the start and end coordinates of the polygon for every scanline. The span is filled by the SLAMs.

Hidden Surface Removal. HSR is not addressed in the published material. It is supposed that the system is implementing some kind of priority based scan conversion algorithm, e.g. painter's algorithm, i.e. the HSR took place already in the geometry processor.

Shading. Halftone patterns. Since this display processor uses only one bit per pixel no colour shading is provided.

Anti-aliasing. None.

5.2.2 Partitioning

This display processor employs a three-level image space partitioning. The main partitioning is in image space; it occurs two times: First, the image is subdivided into groups of 64 scanlines each. In a second step these groups are then split into sets each of which contains 64 × 256 pixels.

Image generation is also partitioned in the functional domain. Vertex sort, producing scanline commands and actual pixel generation are allocated to distinct processors.

The partitioning is therefore characterised as:

$$F(3,1) \circ I(16,64) \circ I(4,64 \times 256)$$

5.2.3 Architecture

Topology. The topology of this display processor is a tree. The root of the tree is formed by the vertex sorter that is connected to the input of the display processor. The scanline processors constitute the first level of the tree. The leaves of the tree are formed by the SLAMs.

Control. This display processor employs a distributed control scheme. Both, scanline processors and SLAMs have their own control unit. The communication between them is not detailed in the literature.

Data types. The scanline processors receive the description of polygons, lines or characters. The primitives are transformed into spans that represented by the x-coordinates of the right and left edge of the span and its y-coordinate. A halftone pattern is provided that is used for filling the span.

The literature does not give details about the representation of the various data.

[7] Y-monotone polygons are characterised by the fact that horizontal lines intersect the boundary of the polyon at most twice. Convex polygons are special cases of monotone polygons. They are monotone in x and y.

Parallelism. All scanline processors receive different primitives from the geometry processor. The commands and data that are generated for each SLAM are different. This characterises a MIMD machine.

Regularity. The display processor is built from two main building blocks ($T = 2$): the scanline processors and the SLAMs. In the following it is assumed that one scanline processor can be integrated on a single chip [**Demetrescu 1985**]. The standard configuration of scanline processors and SLAMs uses 16 scanline processors each of which controls 4 SLAMs. This results in a total chip count of $N = 80$. The regularity is therefore $R = 0.99 \cdot 0.58 = 0.57$.

5.2.4 Performance

The display processor is composed of two different types of chips, scanline processors and SLAMs. These two should be evaluated separately. However, since the scanline processors have been described not very detailed we concentrate on the SLAMs.

The following numbers have been derived from figures given in the literature. (The original paper specified the performance in terms of memory accesses.)

Characteristic parameters. We assume that triangles are processed by the display processor. Since the number of pixels covered by a primitive influence the number of steps to scan convert that polygon the performance of that display processor depends on the properties of the image. Therefore no exact value for α_{sc} can be given. It is estimated for an average size of a primitive of 5,000 pixels and a screen size of 1024 × 1024 pixels.

$$\begin{aligned} t_f &= 1\ s \\ \alpha_{sc} &\approx 0.005 \\ C_{sc} &\approx 50,000 \\ C_F &= 2 \\ C_I &= 64 \end{aligned}$$

This values give a maximum step time of

$$t_{step} \leq 32\ \mu s$$

Primitives per second. [**Demetrescu 1985**] claims that a SLAM can perform approximate 5 million scanline accesses per second. If we make the (conservative) assumption that in average 20 percent of a scanline are covered by a primitive. If we further assume that an object covers 5,000 pixels in average we can estimate the number of primitives processable in one second: 50,000 primitives per second.

Delay time. There is no delay time specified in the literature.

5.3 PROOF

The PROOF system, proposed in [**Schneider 1988a**], is built from three main blocks (figure 4 [**Schneider et al. 1988**]).

The first one, the object processor pipeline (OPP), is made up of a number of *object processors (OP)* storing one object each. Each OP scan converts its object and performs

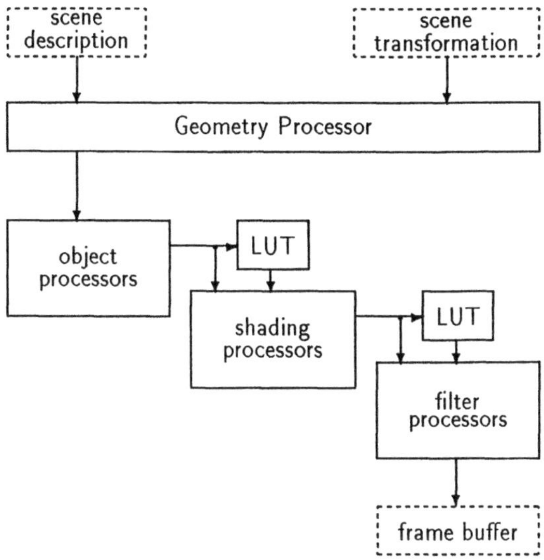

Figure 4: **Block diagram of the PROOF display processor.**

the HSR on a pixel basis. The OPs also interpolate, depending on the configuration of PROOF, colours in RGB or normal vectors.

The next stage in PROOF is a *shading stage* that performs Phong shading. The shading stage is a pipeline of shading processors (SP). Each two SPs are handling one lightsource.

The last stage in PROOF is the *filter stage* that performs the anti-aliasing executing an A-buffer algorithm. There is one processor for each subpixel.

5.3.1 Functional Classification

Primitive Type(s). Triangles and vectors.

Scan Conversion. Inside test. The triangles are described by their bounding lines.

Hidden Surface Removal Depth-buffer (A-buffer).

Shading. Flat and Gouraud shading (OPP only). Phong shading (with shading stage).

Anti-aliasing. A-buffer on a 8×8 subpixel grid.

5.3.2 Partitioning

The PROOF architecture partitions the image generation process in several domains. The first is the functional domain. Each stage carries out another set of functions. The OPP performs scan conversion, HSR and (limited) shading. The shading stage and the filter stage perform illumination and anti-aliasing respectively.

Each stage partitions its task again. The OPP and the shading stage allocate objects (primitives or light sources) to the different processors (object space partitioning). The filter stage provides one processor for each of the 64 subpixels (image space partitioning). This partitioning scheme is represented by

$$F(3,3) \circ [O(50000,1) || O(16,0.5) || I(64,1)]$$

5.3.3 Architecture

Topology. The organisation of PROOF is a pipeline. The OPP and the shading stage are pipelines themselves. The filter stage is constructed from two trees of adders that compute coverage information for every subpixel [**Romanova 1988**]. The output of these trees is fed into an array of subpixel processors that compute the colour of the subpixels.

Control. There is no central control in PROOF. All processing elements, OPs, SPs and FPs, have their own controllers. The different processors are communicating via an asynchronous protocol.

Data types. Information for pixels is propagated through the various pipelines. In the OPP and the shading stage a list of objects is associated with every pixels. The objects in the list are described by an object identifier, depth and normal vector (or colour value) at the current pixel and geometry of the coverage for the current pixel. The filter stage produces a RGB triple. All data items are describe in fixed point format.

Parallelism. In total PROOF is a MIMD machine. The single stages are SIMD processors. All processors in one stage are executing the same instruction (may be delayed due to the pipeline structure) on different data.

Regularity. OPP and shading stage consist only of one type of processing elements, namely the OPs and SPs. Five OPs and two SPs are expected to reside on the same chip. For a system that handles 50,000 objects and 16 light sources 10,000 OP chips and 16 SP chips are needed.

The filter stage uses two types of processing elements; processor slices for the adder trees and subpixel processors. Each of the x-, y- and z-trees will be distributed on three chips. Four subpixel processors will be integrated on one chip. For a resolution of 8×8 subpixels 16 subpixel processor chips are needed.

PROOF is built from four types of chips ($T = 4$) and approximately $N = 10,000$ chips. The population of the chip classes is $n_1 = 10,000$, $n_2 = 16$, $n_3 = 9$ and $n_4 = 16$. From this we obtain $R = 1.0 \cdot 0.50 = 0.50$.

5.3.4 Performance

We assume a screen resolution of 1024×1024 pixels. The average depth complexity may be 2. (The depth complexity directly influences the length of the lists constructed for the A-buffer.)

Characteristic parameters. 1. Object Processors

$$t_f = 33\ ms$$
$$\alpha_{sc} = 2$$

$$C_{sc} = 1$$
$$C_F = 3$$
$$C_I = 1024 \times 1024$$

This gives a step time of $t_{step} = 15.73\ ns$ which is equivalent to a clocking frequency of 63.5 MHz.

2. Shading Processors

$$t_f = 33\ ms$$
$$\alpha_{sc} = 2$$
$$C_{sc} = 1$$
$$C_F = 5$$
$$C_I = 1024 \times 1024$$

This results in a step time of $t_{step} \leq 3.15\ ns$. To enlarge the step time for the individual shading processors, time domain partitioning with up to 16 shading processors in parallel will be employed. This will give a step time of $t_{step} \leq 50\ ns$.

3. Subpixel Processors

$$t_f = 33\ ms/(1024 \times 1024)$$
$$\alpha_{sc} = 2$$
$$C_{sc} = 1$$
$$C_F = 7$$
$$C_I = 1024 \times 1024$$

Since the step time is about 2 ns up to 10 parallel processors will be working in parallel. Thus the step time is enlarged to about 20 ns.

Primitives per second. In [**Schneider 1988a**] a throughput of 50,000 triangles every 33 ms has been announced.

Delay time. The literature gives a worst case delay time of 10.5 ms for a complete image update.

5.4 The Ray-Casting Machine of Kedem & Ellis

The Ray-Casting Machine [**Kedem et al. 1984a**] [**Kedem et al. 1984b**] aims at the scan conversion of objects defined by constructive solid geometry (CSG). In CSG objects are described by set operations, e.g. union, intersection or difference, applied to halfspaces.

The Ray-Casting Machine is built from two types of processors (figure 5). The *primitive classifiers (PC)* calculate the intersection of a ray (a line) with a primitive, i.e. classifies the primitive with respect to the line. Each PC holds one primitive and accepts rays for classification to the stored primitive. The classification returns segments of the ray that are inside or outside of the primitive. The *classification combiners (CC)* combine two classified rays using a certain set operation.

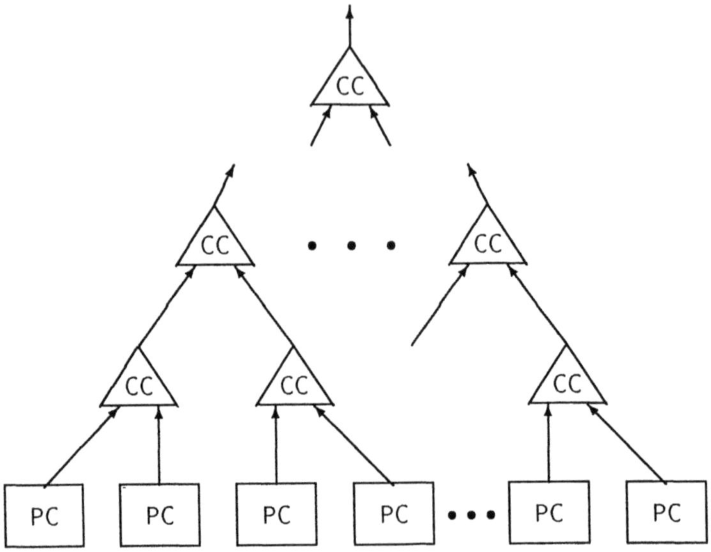

Figure 5: **Block diagram of the Ray-Casting Machine.** PC – Primitive Classifier, CC – Classification Combiners.

5.4.1 Functional Classification

Primitive Type(s). Linear and quadratic halfspaces.

Scan Conversion CSG with ray tracing. The rays into the scene are restricted to be all parallel.

Hidden Surface Removal. Ray tracing.

Shading. None.

Anti-aliasing. None.

Shading and anti-aliasing could be added as a postprocessing step that uses the results of the Ray-Casting Machine.

5.4.2 Partitioning

The Ray-Casting Machine employs an object space and a functional domain partitioning. The object information is distributed to the PCs and CCs. These two types of processors also constitute the functional subdivision: One performs primitive classification the others combine classifications.

The partitioning is characteried as

$$F(2,1) \circ (O(n,1) || O(n \log_2 n, 1))$$

if a maximum of n primitives have to be handled.

5.4.3 Architecture

Topology. The topology is a binary tree. The leaves of the tree are the PCs and the nodes are the CCs. The Ray-Casting Machine can be configured dynamically to the CSG descriptions of different objects. The configuration takes place by loading the CCs such that they implement the required CSG tree.

Control. The control of the Ray-Casting Machine is distributed among all processing elements, PCs and CCs. The processing elements are communicating synchronously using a complete handshake protocol.

Data types. Data moving through the processor tree represent line segments classified as being IN our OUT of the CSG object. These segments are described by integer numbers.

Parallelism. All PCs together form a SIMD computer. The CCs are also working in SIMD mode.

Regularity. The Ray-Casting Machine is built from only one type of chip. Such a chip contains from 4 to 8 slices each of which consists of one PC and 9 CCs. The chip level regularity of the Ray-Casting Machine is therefore $R = 1$.

5.4.4 Performance

The literature does not contain any performance numbers for the Ray-Casting Machine. We can therefore summarise here only those numbers that are a consequence of the architectural features.

Characteristic parameters. 1. Primitive Classifiers

$$
\begin{aligned}
\alpha_{sc} &= 1 \\
C_{sc} &= 1 \\
C_I &= 1024 \times 1024 \\
C_F &= ? \\
t_f &= ?
\end{aligned}
$$

2. Classification Combiners

$$
\begin{aligned}
\alpha_{sc} &= ? \\
C_{sc} &= 2(1) \\
C_I &= 1024 \times 1024 \\
C_F &= ? \\
t_f &= ?
\end{aligned}
$$

5.5 Structured Frame Store System

In [**Jayasinghe et al. 1988**] a system has been presented that can display the contents of the LDF without a frame buffer directly on the output device. The system is based on methods and algorithms developed in [**Akman et al. 1988**] and [**Kuijk et al. 1988**].

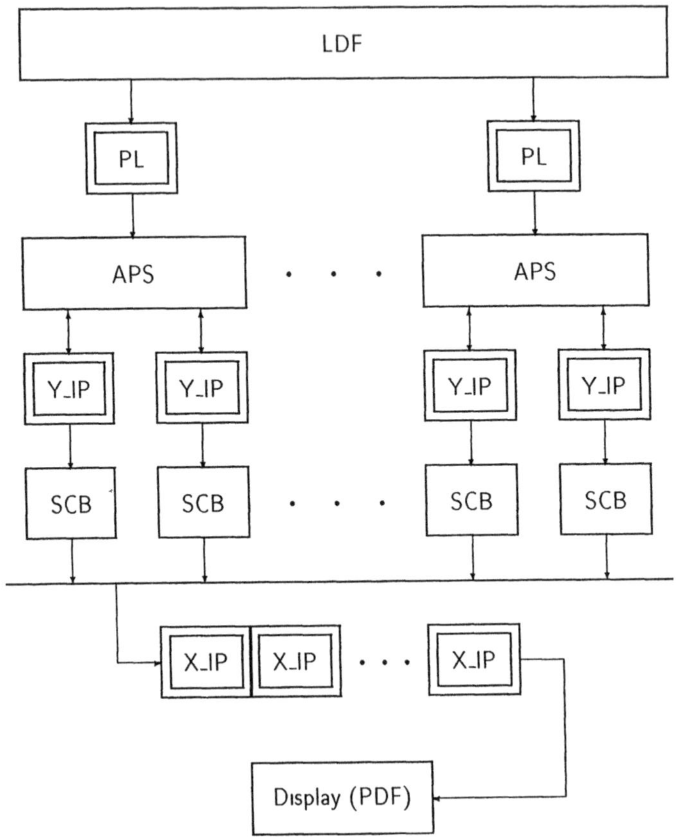

Figure 6: **Display processor of the Structured Frame Store System.** LDF – Linear Display File, APS – Active Pattern Store, SCB – Scanline Command Buffer, X_IP – X Incremental Processor, Y_IP Y – Incremental Processor.

The display controller of that system is constructed from three main building blocks (cf. figure 6 [**Jayasinghe et al. 1988**]). The *Pattern Loader (PL)* receives patterns from the LDF and, if necessary, decomposes them into simpler patterns that can be processed by the rest of the display processor.

These patterns are handed over to *Y Incremental Processors (Y_IP)*. They generate for each pattern a number of scanline commands that are dispatched to pixel processors, called *X Incremental Processors (X_IP)*.

There is one X_IP for every pixel in a scanline. The X_IPs are connected in a pipeline, forming a one-dimensional systolic array. The first X_IP receives a scanline command, executes this command and propagates it to the next X_IP. The task of each X_IP is the incremental computation of pixel intensities.

5.5.1 Functional Classification

Primitive Type(s). The display processor itself accepts patterns [8] as primitives. These general patterns are decomposed into spans because spans are the primitives on which the X_IPs are operating.

Scan Conversion. Contour tracking. The starting and ending pixels of span are incrementally computed by the Y_IPs. The filling is done by the X_IPs.

Hidden Surface Removal. None. The LDF is already preprocessed such that there are only non-overlapping patterns.

Shading. Due to the (restricted) programmability of Y_IPs and X_IPs Wash, flat (constant), Gouraud and Phong shading can be performed. Furthermore, the display processor offers the possibility to generate periodic textures.

Anti-aliasing is also a programmable feature of of the X_IPs. Although not explained in detail, the anti-aliasing technique used seems to be a one-pass technique.

5.5.2 Partitioning

Neglecting the PLs, the display processor performs a two-level partitioning. The first level is a partitioning in the functional domain. The Y_IPs break down the pattern into scanline segments and generate the necessary scanline commands for the X_IPs. Then the X_IPs fill the span.

The second level of partitioning takes place in the time domain for the Y_IPs and in image space for the X_IPs. There are several Y_IPs in order to achieve satisfying performance. Each X_IP handles one pixel in a scanline. This system can therefore be characterised as

$$F(2,1) \circ (T(24 \ldots 245, 1) || I(1024, 1))$$

Furthermore, in [**Jayasinghe et al. 1988**] another partitioning in the time domain is proposed for the X_IPs. If the X_IPs are to slow to perform their task in synchronism with the pixel clock of the output device, each X_IP is replicated in order to enlarge the available time for the individual X_IP.

5.5.3 Architecture

Topology. The topology of the Structured Frame Store's display processor consists of different elements. LDF, PLs, and Y_IPs form a tree structure. The leaves of that tree (the Y_IPs) are working on a common bus. That bus is connected to a systolic array of X_IPs.

Control. The control is distributed across the display processor. The different blocks are communicating via intermediate file structures.

Data types. Input to the PL are patterns on a fairly high level. The exact representation of these patterns is not known to the author. These patterns are transformed into polygons described by their geometry and associated colour information. The polygons are decomposed into single spans by the Y_IPs. The X_IPs translate the spans into pixel information, e.g. RGB values.

[8] In [**Kuijk et al. 1988**] patterns are defined as flat regions in continuous 3D. They may be disconnected, concave, with holes or with islands.

Parallelism. There are different kinds of parallelism in this display processor. The PLs are working in parallel on different primitives that may have different types. They are therefore forming a MIMD processor. The Y_IPs are all executing the same command on different primitives which characterises them as a SIMD architecture. This is also the case for the X_IPs that are all executing the same scanline commands.

Regularity. It had not been fixed in the literature how the PLs, Y_IPs and X_IPs will be implemented. It is therefore not possible to determine the regularity of the system.

5.5.4 Performance

We will look only at the X_IP and the Y_IP because the PL is performing functions that we originally associated with the geometry processor. Since these two blocks are differing they are considered separately.

We will assume a screen of 1024 × 1024 pixels. The primitives should have an average height of 70 pixels. (Some of the values of the following parameters stem from **[Jayasinghe 1989]**.)

Characteristic parameters. 1. Y_IP

$$
\begin{aligned}
t_f &= 20\ ms \\
\alpha_{sc} &= 0.07 \\
C_{sc} &= 244,000s^{-1} \cdot 20\ ms \ = \ 4880 \\
C_F &= 60 \\
C_I &= 1024
\end{aligned}
$$

This results in $t_{step} = 0.95\ ns$ for one (!) Y_IP. It is proposed that for a screen with 1024 × 1024 pixels and Gouraud shading ≈ 122 Y_IPs are used which reduces the functional complexity to $C_F = 60/122 \approx 0.5$. For the new step time we obtain $t_{step} = 0.116\ \mu s$ (≡ 8.6 MHz).

2. X_IP

$$
\begin{aligned}
t_f &= 20\ ms/1024 = 20\ \mu s \\
\alpha_{sc} &= 1 \\
C_{sc} &= 4880/70 = 70 \\
C_F &= 4 \\
C_I &= 1
\end{aligned}
$$

The step time for the X_IP is $t_{step} = 71.4\ ns$. This is equivalent to a clocking frequency of 14 MHz.

Primitives per second. In **[Jayasinghe et al. 1988]** it is claimed that 244,000 Gouraud shaded quadrilaterals of height 70 can be rendered per second.

Delay time. The delay time of this display processor is one frame time (20 ms).

6 Possible Novel Display Processor Architectures

The investigation and classification of several display processors revealed that there are some "white spots" on the map of possible architectures. Two of them that deserve a closer look are the following: Tree structured display processors that partition in image space and display processors on the basis of 2D-arrays of processing elements.

We want to suggest an architecture that closes the first of these gaps: Quadtrees and octrees subdivide the scene or the screen recursively until certain criteria are met. Subdivision of an image into subimages (cells) stops if

- A cell is empty, i.e. contains no object.
- A cell contains exactly one object.
- A predefined level of subdivision has been reached.

This approach offers the possibility of a rendering process that quickly produces an image with a coarse resolution. Afterwards the image is refined further and further until the final accuracy has been reached. Such a behavior is desirable in an environment where response time dominates the desire for image quality.

An architecture for such a display processor could be a tree structure of processors that reflects the structure of the quadtree or octree. A problem in that approach is that it is impossible to provide a tree structure that contains a processor for every possible cell. Therefore, the tree has to be dynamically configurable. It must also be able to detect and reasonably react to overload conditions.

7 Conclusion

The presented taxonomy is mainly founded on a model for the display processor. The model considers two aspects of the display architecture, function and timing. The classifications for *function*, *partitioning* and *performance* have been derived from them. The last distinguishing feature, the display processor's architecture, could not be based on that model, it has been defined mostly intuitively. However, there is a close relation between the topology of an architecture and the employed partitioning scheme. Improvements of the taxonomy can be achieved by refining the model. Especially, an inclusion of architectural aspects into the model seems promising.

The second part of the paper applied the taxonomy to existing display processors. Several times we were unable to completely classify a display processor. That was due to a lack of information in the published material. The taxonomy can serve as a guideline to identify and select the minimum information that is needed to describe a display processor. This would mean to make best use of a taxonomy.

References

[**Abram et al. 1986**] Gregory D. Abram and Henry Fuchs: *VLSI Architectures for Computer Graphics*. In G. Enderle, editor, *Advances in Computer Graphics I*, pages 6–21, Springer-Verlag, Berlin, Heidelberg, New York, Tokyo, 1986.

[**Akman et al. 1988**] Varol Akman et al.: *A Vector-like Architecture for Raster Graphics*. In A.A.M. Kuijk and Wolfgang Straßer, editors, *Advances in Computer Graphics Hardware II*, pages 137–154, Eurographics, Springer-Verlag, 1988.

[Blinn 1989a] James F. Blinn: *Return of the Jaggy.* IEEE Computer Graphics & Applications, 82–89, March 1989.

[Blinn 1989b] James F. Blinn: *What We Nee Around Here Is More Aliasing.* IEEE Computer Graphics & Applications, 75–79, January 1989.

[Carlbom 1980] Ingrid B. Carlbom: *System Architecture for High-Performance Vector Graphics.* PhD thesis, Dept. of Computer Science, Brown University, Providence, R.I., 1980.

[Carpenter 1984] Loren Carpenter: *The A-buffer, an Antialiased Hidden Surface Method.* Computer Graphics, 18(3):103–108, July 1984.

[Clark 1982] James H. Clark: *The Geometry Engine: A VLSI Geometry System for Graphics.* Computer Graphics, 16(3):127–133, July 1982.

[Claussen 1988] Ute Claussen: *Beleuchtungsmodell und Beleuchtungsalgorithmen in der Graphischen Datenverarbeitung.* Forschungsbericht WSI-GRIS 88-3, Wilhelm-Schickard-Institut für Informatik, Universität Tübingen, WSI-GRIS, Auf der Morgenstelle 10/C9, D-7400 Tübingen, W-Germany, 1988.

[Demetrescu 1985] Stefan Demetrescu: *High Speed Image Rasterization Using Scan Line Access Memories.* In Henry Fuchs, editor, *Proceedings of the Chapel Hill Conference on VLSI,* pages 221–243, Computer Science Press Inc., 1803 Research Blvd., Rockville, Maryland 20850, 1985.

[Dew et al. 1985] Peter M. Dew et al.: *Systolic Array Architectures for High Performance CAD/CAM Workstations.* In Rae A. Earnshaw, editor, *Fundamental Algorithms for Computer Graphics,* pages 659–694, NATO ASI, Springer-Verlag, Berlin, Heidelberg, New York, Tokyo, 1985.

[Fuchs 1988] Henry Fuchs: *An Introduction to Pixel-planes and other VLSI-intensive Graphics Systems.* In R.A. Earnshaw, editor, *Theoretical Foundations of Computer Graphics and CAD,* chapter 6, pages 675–688, Springer-Verlag, 1988.

[Fuchs et al. 1982] Henry Fuchs et al.: *Developing Pixel-Planes. A Smart Memory-Based Raster Graphics System.* In *Proceedings of the Conference on Advanced Research in VLSI, M.I.T.,* pages 137–146, January 1982.

[Fuchs et al. 1985] Henry Fuchs et al.: *Fast Spheres, Shadows, Textures, Transparencies, and Image Enhancements in Pixel-Planes.* Computer Graphics, 19(3):111–120, July 1985.

[Fuchs et al. 1988] Henry Fuchs et al.: *Coarse-Grain and Fine-Grain Parallelism in the Next Generation Pixel-planes Graphics System.* In *Proceedings of the International Conference and Exhibition on Parallel Processing for Computer Vision and Display,* University of Leeds, UK, January 1988.

[Ghazanfarpour et al. 1987] Djamchid Ghazanfarpour and Bernard Peroche: *A Fast Antialiasing Method with a Z-Buffer.* In G. Maréchel, editor, *Eurographics '87,* pages 503–512, Eurographics, Elsevier Science Publishers B.V. (North Holland), 1987.

[**Giloi 1983**] W.K. Giloi: *Towards a Taxonomy of Computer Architecture Based on the Machine Data Type View.* In *The 10th Annual Symposium on Computer Architecture*, pages 6–15, ACM, 1983.

[**Hu et al. 1985**] Mei-Cheng Hu and James D. Foley: *Parallel Processing Approaches to Hidden-Surface Removal in Image Space.* Computer & Graphics, 9(3):303–317, 1985.

[**Jayasinghe 1989**] J.A.K.S. Jayasinghe: *Personal Communications.* July 1989. —.

[**Jayasinghe et al. 1988**] J.A.K.S. Jayasinghe et al.: *A Display Controller for a Structured Frame Store System.* In A.A.M. Kuijk, editor, *Third Eurographics Workshop on Graphics Hardware*, Eurographics, 1988. The proceedings will be published in the book "Advances in Graphics Hardware III" by Springer-Verlag in 1989.

[**Kedem et al. 1984a**] Gershon Kedem and John L. Ellis: *Computer Structures for Curve-Solid Classification in Geometric Modeling.* Technical Report TR84-137, Rochester Institute of Technology, Microelectronics Center of North Carolina, September 1984.

[**Kedem et al. 1984b**] Gershon Kedem and John L. Ellis: *The Raycastig Machine.* In *The Proceedings of the 1984 International Conference on Computer Design*, pages 533–538, IEEE, 1984.

[**Kilgour 1981**] A.C. Kilgour: *A Hierarchical Model of a Graphics System.* Computer Graphics, 15(1):35–47, April 1981.

[**Kilgour 1985**] Alistair C. Kilgour: *Parallel Architectures for High Performance Graphics Systems.* In Rae A. Earnshaw, editor, *Fundamental Algorithms for Computer Graphics*, pages 695–703, NATO ASI, Springer-Verlag, Berlin, Heidelberg, New York, Tokyo, 1985.

[**Kuijk et al. 1988**] A.A.M. Kuijk et al.: *An Exact Incremental Hidden Surface Removal Algorithm.* In Fons Kuijk and Wolfgang Straßer, editors, *Advances in Computer Graphics Hardware II*, pages 21–38, Eurographics, Springer-Verlag, 1988.

[**Lakshminarasimhan et al. 1989**] A.L. Lakshminarasimhan and Mandayam Srivas: *A Framework for Functional Specification and Transformation of Hidden Surface Elimination Algorithms.* Computer Graphics Forum, 8(2):75–98, June 1989.

[**Latham 1985**] Roy W. Latham: *Image Generator Architectures and Features.* In Roy Latham, editor, *Course notes for high performance image generation systems*, chapter 7, —, Link Flight Simulator Division - The Singer Company - Sunnyvale, CA, July 1985. Reprinted from the 5th Interservice/Industry Training Equipment Conference, 1983.

[**Parke 1980**] Frederic I. Parke: *Simulation and Expected Performance Analysis of Multiple Processor Z-Buffer Systems.* Computer Graphics, 13(3):48–56, July 1980.

[**Pineda 1988**] Juan Pineda: *A Parallel Algorithm for Polygon Rasterization.* Computer Graphics, 22(4):17–20, August 1988.

[**Porter et al. 1984**] Thomas Porter and Tom Duff: *Compositing Digital Images.* Computer Graphics, 18(3):253–259, July 1984.

[**Poulton et al. 1985a**] John Poulton et al.: *Pixel-Planes : Building a VLSI-Based Graphic System.* In Henry Fuchs, editor, *Chapel Hill Conference on Very Large Scale Integration*, pages 35–60, Computer Science Press, Inc., 1803 Research Blvd., Rockville, Maryland, 1985.

[**Poulton et al. 1985b**] John Poulton et al.: *Pixel-planes graphic engine.* In Neil H. E. Weste and Kamran Eshraghian, editors, *Principles of CMOS VLSI Design – A Systems Perspective*, pages 448–480, Addison-Wesley, Reading, Mass., 1985.

[**Pulleyblank et al. 1987**] Ron Pulleyblank and John Kapenga: *The Feasibility of a VLSI Chip for Ray Tracing Bicubic Patches.* IEEE Computer Graphics & Applications, 7(3):33–44, March 1987.

[**Reghbati et al. 1988**] Hassan K. Reghbati and Anson Y.C. Lee: *Computer Graphics Hardware: Image Generation and Display.* IEEE Computer Society, 1988.

[**Robertson 1988**] Philip K. Robertson: *Visualizing Color Gamuts: A User Interface for the Effective Use of Perceptual Color Spaces in Data Displays.* IEEE Computer Graphics & Applications, 50–64, September 1988.

[**Romanova 1988**] Claudia Romanova: *Effizientes Anti-Aliasing für die Bilderzeugung auf Rastersichtgeräten.* In W. Barth, editor, *Visualisierungstechniken und Algorithmen*, pages 109–118, Gesellschaft für Informatik, Österreichische Computer Gesellschaft, Springer Verlag, September 1988.

[**Schneider 1988a**] Bengt-Olaf Schneider: *A Processor for an Object-Oriented Rendering System.* Computer Graphics Forum, 7:301–310, 1988.

[**Schneider 1988b**] Bengt-Olaf Schneider: *Ray Tracing Rational B-Spline Patches in VLSI.* In A.A.M. Kuijk and Wolfgang Straßer, editors, *Advances in Graphics Hardware II*, Eurographics, Springer-Verlag, 1988.

[**Schneider et al. 1988**] Bengt-Olaf Schneider and Ute Claussen: *PROOF: An Architecture for Rendering in Object Space.* In *Third Eurographics Workshop on Graphics Hardware*, Eurographics, 1988. The proceedings will be published in the book "Advances in Graphics Hardware III" by Springer-Verlag in 1989.

[**Sutherland et al. 1974**] Ivan E. Sutherland et al.: *A Characterization of Ten Hidden-Surface Algorithms.* ACM Computing Surveys, 6(1):1–55, March 1974.

A Hardware Algorithm for Fast Realistic Image Synthesis

A. C. Yilmaz, S. Hagestein, E. Deprettere and P. Dewilde

A VLSI oriented algorithm, for the implementation of a generalized two-pass radiosity method is presented. The method allows any reflection behavior, varying from purely diffuse to perfect mirroring. Moreover, objects may be defined in terms of curved (Bezier) surfaces. All computations in the pre- and postprocess are similar and ray-tracing based, consequently a single architecture can be devised for both passes. This architecture, when built on ray-rotating and ray-tracing pipelined processors such as Cordics, results in a very high throughput VLSI implementation of the proposed generalized two-pass procedure.

1. Introduction

In recent years, new methods for realistic image rendering have been presented. Ray-tracing and radiosity methods appear to be most suitable for realistic image synthesis. Ray-tracing methods are conceptually simple and especially suited to model transparency and near mirroring reflections. The radiosity method, based on energy transport and conservation principles, models correctly the diffuse interreflection within an environment. It has been successfully applied in thermal heat transfer simulations [18,19] and was reintroduced in computer graphics by Goral et al. [9]. The method is computationally intensive since it requires the computation of the entries - the so-called form-factors - in a large matrix and the iterative solution of a set of linear equations involving this matrix. The computation of form-factors can be simplified by projecting the patch onto a hemicube [5], though this leads to an inhomogeneous sampling of the environment above the cube. We will use the original hemisphere instead of the hemicube. The accuracy of the diffuse intensity has been improved by an adaptive patch subdivision method [4] in which the rendering of shadows and penumbras are accounted for. The implementation proposed in [4] uses a hierarchical data structure requiring adjacency relationship within the environment. An alternative approach will be presented in this paper. A first attempt to extend the method to include non-diffuse surfaces was proposed by Immel et al. in [11]. Since this method computes a complete view-independent intensity distribution over all patches, an unacceptable high cube resolution is required.

A practical method to deal with complex environments was introduced by Wallace et al. in [20]. They proposed a two-pass approach consisting of a view-independent preprocess, based on the radiosity method, and a view-dependent postprocess, based on ray-tracing. In the preprocess, form-factors are computed by using a hemicube and a projection technique. In [20], so called "extra form-factors" representing the extra amount of reflected energy via perfect mirrors are taken into account. In [15] the restriction to perfect mirrors was removed by using a procedural iteration to approximate differently defined extra form-factors. Although this approach can reduce the processing time as compared to the approach presented by Immel [11], the following drawbacks can not be ignored. Firstly, in case of high intensity gradients, a lot of memory is needed to store the visibility information for each non-diffuse patch. Secondly, as with [4], a very high hemicube resolution is required to render the directional intensity gradient accurately.

In this paper we present a novel two-pass procedure for the visualizing of complex environments. The procedure is a modification to the approach in [20], in the sense that it is both generalized and specialized. It is more general, not only because it allows any reflection behavior, from purely diffuse to perfect mirroring, but also because it accepts curved (Bezier) surface descriptions. The procedure is specialized in that it is optimized in terms of hardware, in particular high speed ASIC implementation.

The first pass (*preprocess*): In this pass we compute form-factors by using a hemisphere and ray-tracing. A set of rays is used to define a reference hemisphere. Each hemisphere ray refers to a precomputed delta form-factor specifying the incoming energy within a small solid angle. This energy arrives from a diffuse patch, directly or indirectly via specular reflections. For each diffuse patch the rays are rotated, by using Cordic arithmetic, and cast. If the ray intersects a diffuse patch directly, the corresponding delta form-factor is added to the form-factor between the two involving patches. If the ray intersects the same diffuse patch indirectly, via specular reflections, a weighted delta form-factor is added to the form-factor. So, we compute correct intensities on diffuse patches by taking multiple specular reflections into account, but without computing specular patch intensities themselves. The size of the form-factor matrix is independent of the number of specular patches. The use of a hemisphere makes an efficient storage of the frustum (or the cone) weight-factors possible, so that reflections depending on a varying solid angle and a reflection power (see section 3.3) can be easily modeled. The storage problem of multiple reflection and transmission within varying frustums has been overcome by applying a backtracking procedure. All intersections are computed by using a subdivision and a bounding box algorithm. By using a bounding box algorithm, patches are no longer restricted to planar polygons but Bezier forms can be processed as well. Once the form-factor matrix is constructed the diffuse patch intensities are obtained by applying a Gauss-Seidel iterative matrix equation solver [3].

Instead of the above mentioned adaptive patch subdivision method we apply an "adaptive patch sampling" method, which avoids the need to using a hierarchical data structure requiring the adjacency information. The sample points on patches are controlled by parametric (u,v) coordinates.

The second pass (postprocess): In this pass we cast rays from a given eye position through the screen pixels and compute intersections with the corresponding patches. If the intersected patch is diffuse then the diffuse intensity component of the screen pixel is obtained by bi-linear interpolation using intensity values computed in the first pass. If the patch is specular then the specular intensity component is computed by shooting rays confined to a frustum.

It should be noted that shooting rays from a hemisphere in the preprocess similar to shooting rays through the displaying screen in the postprocess and the computation of extra light on diffuse patches, caused by multiple specular reflections, in the preprocess is similar to the specular intensity computation in the postprocess. Moreover, because both passes of the algorithm are ray-tracing based, a single architecture is naturally used to implement both passes. Since all rays computations are both similar and independent of each other, the whole procedure is optimally tailored towards a very fast SIMD/wavefront-type architecture.

The outline of this paper is as follows. In section 2 we give a concise review of the radiosity method. In section 3 we will show how light energy transfer can be modeled by defining, what we called, direct and indirect form-factors. Also in this section we will show how these form-factors can be computed by approximating patch to hemisphere projections. Adaptive patch sampling is described in section 4. In section 5, attention will be paid to the postprocess, including the view dependent specular intensity computation stage and the rendering stage. In section 6, a preliminary conceptual hardware version is given. In section 7 it is shown how Bezier patches are processed. Finally some statistics and software simulation results are given in section 8.

2. The Illumination Equation

Kajiya [12] and Immel [11] have shown that the intensity of light reflected by a surface obeys the following illumination equation:

$$I_{out}(\theta_{out}) = E(\theta_{out}) + \int_{\Omega} \rho''(\theta_{out},\theta_{in})\, I_{in}(\theta_{in}) \cos\theta \, d\omega, \qquad (2.1)$$

where

θ_{in} = the incoming direction, described by polar angle θ and azimuthal angle ϕ.
θ = the angle between the incoming direction and the patch normal.
$I_{out}(\theta_{out})$ = the intensity in outgoing direction θ_{out}.
$E(\theta_{out})$ = the outgoing intensity in direction θ_{out} due to emission.
$\rho''(\theta_{out},\theta_{in})$ = bidirectional reflectance/transmittance coefficient.
$I_{in}(\theta_{in})$ = the incoming intensity in direction θ_{in}.
Ω = the sphere of incoming directions
$d\omega$ = the solid angle through which I_{in} arrives.

In a number of papers [16,7,10] it was shown that with very little loss of accuracy, the bidirectional reflection coefficient can be approximated as the sum of a view-independent portion $k_d\rho_d$ and a view-dependent portion $k_S\rho_S(\theta_{out},\theta_{in})$. Further separation of both portions into a reflected and transmitted part yields:

$$\rho''(\theta_{out},\theta_{in}) = k_d\rho_d + k_y\rho_y + k_r\rho_r(\theta_{out},\theta_{in}) + k_t\rho_t(\theta_{out},\theta_{in}), \qquad (2.2)$$

where

k_d = diffuse fraction of reflected light energy
k_y = diffuse fraction of transmitted light energy
k_r = specular fraction of reflected light energy
k_t = specular fraction of transmitted light energy,

with $k_d + k_y + k_r + k_t \leq 1$.

Reflection models for surfaces have been described by Phong [16], Cook et al. [7] and Hall et al. [10]. Because of its simplicity, Phongs' model (see section 3.3) is the commonly used one. It is a reasonable approximation of the specular reflection process [7], based on empirical observations. Substituting (2.2) in (2.1) gives us:

$$I_{out}(\theta_{out}) = E(\theta_{out}) + I_{d,out} + I_{y,out} + I_{r,out}(\theta_{out}) + I_{t,out}(\theta_{out}), \qquad (2.3)$$

where

$$I_{d,out} = k_d\rho_d \int_{\Omega} I_{in}(\theta_{in}) cos(\theta) d\omega, \quad I_{r,out}(\theta_{out}) = k_r \int_{\Omega} \rho_r(\theta_{out},\theta_{in}) I_{in}(\theta_{in}) \cos(\theta) d\omega,$$

$$I_{y,out} = k_y\rho_y \int_{\Omega} I_{in}(\theta_{in}) cos(\theta) d\omega, \quad I_{t,out}(\theta_{out}) = k_t \int_{\Omega} \rho_t(\theta_{out},\theta_{in}) I_{in}(\theta_{in}) \cos(\theta) d\omega.$$

Although we may separate the reflectance into diffuse and specular terms, we can not compute diffuse intensity components $I_{d,out}$ and $I_{y,out}$ independently of the specular intensity components $I_{r,out}(\theta_{out})$ and $I_{t,out}(\theta_{out})$. Indeed, an incoming intensity $I(\theta_{in})$ is just an outgoing intensity $I_{out}(\theta_{out})$ of another patch which in turn can contain both diffuse and specular components. In the preprocess, the effect of incoming specular intensities is accounted for by defining what we called indirect form-factors, which will be illustrated in section 3.3.

Let us assume that all diffuse surfaces in an environment are subdivided into patches, in such a way that each patch exhibits a constant intensity. Then for each diffuse patch pair the geometrical relation (form-factor) representing the fraction of energy leaving a patch and landing on another patch can be drawn up. The form-factor between a differential area dA_i of patch i and another patch j, see figure 1, is defined [5] as:

$$D_{dA_iA_j} = \int_{A_j} \frac{\cos\phi_i \cos\phi_j}{\pi r^2} HID(i,j)\, dA_j, \tag{2.4}$$

where

$HID(i,j) = 1$ if patch i sees patch j and 0 otherwise.

The form-factor between two diffuse patches i and j is found by taking the area average of (2.4) and is given by:

$$D_{A_iA_j} = \frac{1}{A_i} \int_{A_i} \int_{A_j} \frac{\cos\phi_i \cos\phi_j}{\pi r^2} HID(i,j)\, dA_j dA_i \tag{2.5}$$

If the distance between the two patches is large enough, then the integrand of the inner integral in (2.5) remains almost constant [5], hence (2.4) can be used instead of (2.5) to compute the form-factor. Once the form-factors between all diffuse patches have been determined, a set of linear equations can be deduced. In matrix form they look as follows:

$$\mathbf{F}_{R,G,B} \cdot \mathbf{I}_{R,G,B} = \mathbf{E}_{R,G,B}, \tag{2.6}$$

where

$\mathbf{I}_{R,G,B}$ is the vector of intensities,
$\mathbf{E}_{R,G,B}$ is the vector of light source intensities,
$\mathbf{F}_{R,G,B}$ is the form-factor matrix.

Each entry $\mathbf{F}_{R,G,B}$ consists of the product of a form-factor and a diffusion factor ($k_d\rho_d$ or $k_y\rho_y$) of the corresponding patch pair. This formula represents the radiant intensity equation of an environment. The matrix equation can be advantageously solved by using a Gauss-Seidel iteration method.

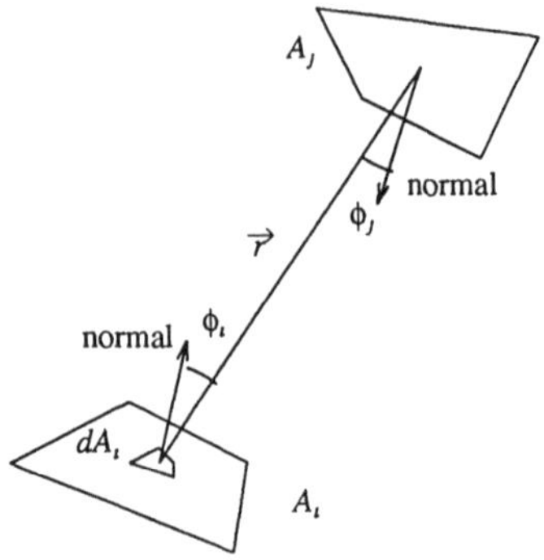

Figure 1. Form-factor geometry.

3. Hemisphere and Form-factors

3.1 Form-factors Between Diffuse Patches: Direct Contributions

As stated in the previous section, a form-factor represents the fraction of energy leaving one surface and landing on another. The *direct* form-factor specifies the direct energy transfer between two diffuse patches and is proportional to the projection of one patch onto a hemisphere placed on the other patch (figure 2a). The direct form-factor from diffuse patch i to diffuse patch j can be defined as:

$$D_{ij} = \frac{1}{\pi}\int_{\xi}\int_{\eta} G'(\eta,\xi)\cos\xi\, d\eta\, dxi,$$

which is a spherical coordinate equation to (2.4), for r=1.

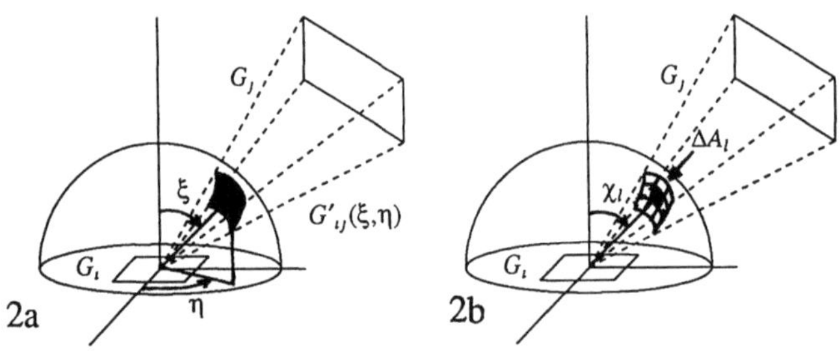

Figure 2. Hemisphere projection (a) and hemisphere discretization (b).

Discretization of the hemisphere into constant differential areas ΔA, figure 2b, yields:

$$D_{ij} = \lim_{\Delta A_l \to 0} \frac{1}{\pi} \sum_l \Delta A_l \cos\chi_l \approx \frac{\Delta A}{\pi} \sum_l \cos\chi_l = K^{-1} \sum_l \cos\chi_l,$$

for ΔA sufficiently small. The normalization factor $K = \Delta A^{-1}\pi$ results from the fact that the fraction of energy emitted within the entire hemisphere equals 1. Hence,

$$K = \frac{\pi}{\Delta A} = \sum_{k=1}^{N} \cos\chi_k,$$

where N is the total number of differential areas on the hemisphere. Each differential area ΔA_l, angle, χ_l thus has its own contribution to the form-factor, called delta form-factor $\delta D_{ij} = K^{-1} \cos\chi_l$.

3.2 Hemisphere Discretization

Although a uniform discretization of the surface of the hemisphere is not easy and does not lead to a uniform distribution of rays defining the hemisphere, a simple and efficient, yet effective method is the following. The rays are such directed that they intersect the hemisphere on circles parallel to the xy-plane with polar angle $j\Delta\theta$, $\Delta\theta$= constant and j= 0,1,...,N, see figure 3.

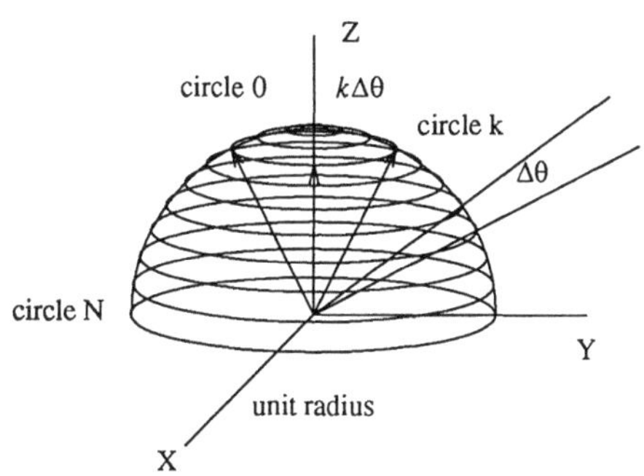

Figure 3. Hemisphere circles

Let the number of rays intersecting circle N, which lies in the XY plane, be equal to M. The distance between two adjacent rays within this circle is then $\Delta x = 2\pi M^{-1}$. We can also make the distance between two adjacent circles equal to Δx, by subdividing the hemisphere into $M/4$ circles, which makes $\Delta\theta = 2\pi M^{-1}$. Two adjacent rays within an arbitrary circle j with angle $j\Delta\theta$ (see figure 3) should then also be spaced by that same distance Δx. The number of rays through circle j must therefore be approximately equal to

$$N_j = \text{round}(M \sin(j\Delta\theta)). \tag{3.1}$$

The distance Δx_j between two adjacent rays through circle j is then

$$\Delta x_j = \frac{2\pi \sin(j\Delta\theta)}{\text{round}(M(\sin(j\Delta\theta))}.$$

Once the rays that define the quantized hemisphere are determined, their individual contribution is computed and stored in a look-up table. Notice that all rays lying on a single circle have the same delta form-factor. Hence the number of contributions that has to be stored equals the number of circles on the hemisphere. The same hemisphere can also be used to define the frustums that are needed to model directional reflection functions. This will be exposed in the next section.

Remark: By rounding off the number of rays through circle j, $M\sin(j\Delta\theta)$, to an integer that is a multiple of four, only one quarter of the entire hemisphere has to be stored. The error induced by this quantization becomes smaller as the number of rays increases.

3.3 Indirect Contributions to Form-factors

In this section we show how extra form-factors can be computed by taking multiple specular reflections into account, as suggested by Wallace et al. in [20], without computing specular patch intensities. In [20] the form-factor was defined as the quantity specifying the total energy transfer between two diffuse patches. However, this transfer can take place via a strictly direct route (if there are no obstacles) as well as via specular reflections, see figure 4. Consequently, we may write the total received energy of a patch as the sum of the energy received via direct and indirect, specular reflective, paths (see figure 4). These are represented by a direct (D_{ij}) and a indirect (S_{ij}) form-factor contributions respectively, thus:

$$F_{ij} = D_{ij} + S_{ij} = D_{ij} + \sum_{p=1}^{path} S_{ij}^{p}$$

where

F_{ij} is the total form-factor.
D_{ij} stands for the direct energy transfer from diffuse patch i to diffuse patch j,
S_{ij} stands for the energy transfer via (multiple) specular reflection(s),
path is the total number of indirect routes along which energy transfer takes place.

The computation of D_{ij} has been discussed in section 3.1. Here we shall put emphasis on S_{ij}, and we will confine ourselves to a practical example. Suppose that patches 3 and 4 reflect light according to Phong's reflectance function [16] which is given by ($cos(\alpha)^n$), where n is the reflection power and α is an angle between the actual reflection direction and the mirror direction. The specular

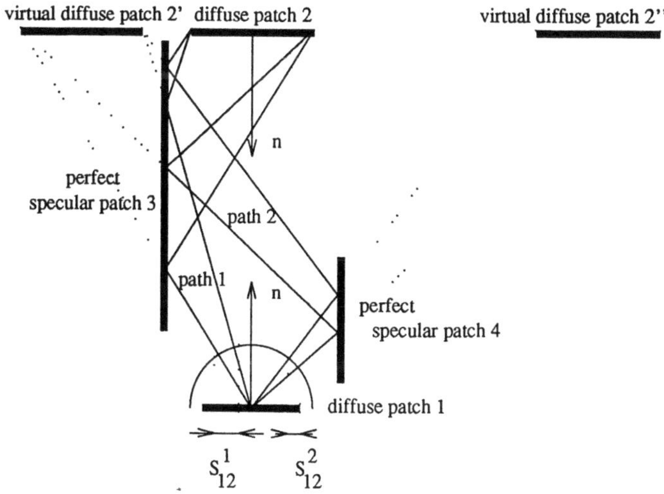

Figure 4. Fractions of the form-factor via perfect specular patches.

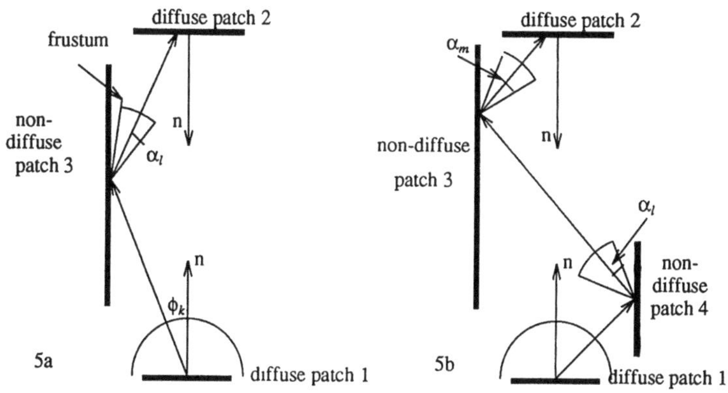

Figure 5. Fractions of the form-factor via non-diffuse patches.

reflectance of a patch may vary from almost diffusely ($n>1$) to perfect mirroring ($n \rightarrow \infty$). To model such a reflectance, a so-called view-frustum is used which in essence is a cone around the z-axis of the hemisphere in figure 3 and defined by a corresponding number of rays from the hemisphere. The frustum size can be adapted to a particular patch reflection low by choosing an appropriate number of rays from the hemisphere. A reflection frustum is directed in the correct position by rotating the rays confined to the frustum. For a given reflection power n, each frustum ray is weighted by a factor, which depends on the reflection angle α.

The indirect form-factors, e.g. between patch 1 and 2 in figure 5a, can be formulated by simply tracing a ray from the hemisphere (patch 1) until it arrives at diffuse patch 2 after a number of specular reflections. For each hemisphere ray intersecting specular patch 3, a frustum of rays will be cast in the reflection direction. Then the contribution of a hemisphere ray to the indirect form-factor S_{12} in path 1, will be multiplied by a factor whose magnitude depends on the position of the ray within the frustum and the number of frustum rays that hits patch 2. This factor is equal to 1 if all frustum rays hit patch 2 and 0 if all rays miss it. Assuming that patch 3 has a reflection power n_3, the additional form-factor between patches 1 and 2 via one reflection (path 1) as shown in figure 5a, is given by:

$$S_{12}^1 = K^{-1} \sum_{k=1}^{y} \cos(\phi_k) \left\{ k_{s3} K'^{-1} \sum_{l=1}^{y_k'} (\cos(\alpha_l))^{n_3} \right\}$$

where

y is the number of hemisphere rays that hit patch 3,
y_k' is the number of frustum rays that hit patch 2,
K'^{-1} is the normalization factor of the frustum.

K'^{-1} results from the observation that the incoming light energy is equal to the scattered light energy in the outgoing direction (if there is no absorption). Similarly, the indirect form-factor via two reflections (path 2), see figure 5b, can be written as:

$$S_{12}^2 = K^{-1} \sum_{k=1}^{z} \cos(\phi_k) \left\{ k_{s4} K'^{-1} \sum_{l=1}^{z_k'} (\cos(\alpha_l))^{n_4} \left\{ k_{s3} K''^{-1} \sum_{m=1}^{z_{kl}''} (\cos(\alpha_m))^{n_3} \right\} \right\}$$

where

z is the number of hemisphere rays that hit patch 4,
z_k' is the number of frustum rays on patch 4 that hit patch 2 via patch 3,
z_{kl}'' is the number of frustum rays on patch 3 that hit patch 2,
K'^{-1} is patch 4 frustum normalization factor.
K''^{-1} is patch 3 frustum normalization factor.

As mentioned before, we can see frustums as part of the hemisphere. Now we shall discuss the computation and the storage of the weight-factors that are assigned to the frustum rays. Consider the hemisphere in figure 3, with N circles on its surface: j=0,1,...,$N-1$. (The circle with number N is not relevant because of the zero contribution.) Then it is possible to construct $N-1$ (j=0,1,...,$N-2$) frustums with different sizes. Observing that all rays lying on a single circle have equal contributions, only one weight-factor for each circle has to be computed. Hence the storage capacity can be reduced considerably.

The circle 0, which has a zero radius, is represented by one ray, i.e. from (3.1), N_0=1. For circle j the number of rays N_j is given by (3.1). The weight-factors are computed by discretizing the Phong model in accordance with the hemisphere discretization. The reflection power and angle then become $n = n_k$ (k=0,1,...,$N-2$) and $\alpha = j\Delta\theta$ (j=0,1,...,$N-2$), respectively. The weight-factor for circle j within a frustum with size k is then:

$$W_{jk} = \frac{(\cos(j\Delta\theta))^{n_k}}{\sum\limits_{i=0}^{k} (\cos(i\Delta\theta))^{n_k} \cdot N_i}$$

where $0 \le k \le N-2$ and $0 \le j \le k$.

The denominator provides for the normalization. Using this formula, a weight-factor table can be set up, as follows:

frustum size k / circle j	0	1	2	$N-2$	$N-1$
0	1	W_{01}	$W_{0,2}$	$W_{0,N-2}$	$W_{0,N-1}$
1	0	$W_{1,1}$	$W_{1,2}$	$W_{1,N-2}$	$W_{1,N-1}$
2	0	0	$W_{2,2}$	$W_{2,N-2}$	$W_{2,N-1}$
$N-2$	0	0	0	$W_{N-2,N-2}$	$W_{N-2,N-1}$
$N-1$	0	0	0	0	$W_{N-1,N-1}$

Remark: The column k=$N-1$ represents the contribution of the hemisphere rays to the form-factor, which is a limiting case where the frustum (cone) becomes the hemisphere itself.

3.4 Aproximation of Projections by Ray-Rotating and Ray-Tracing

The projection of area G_j (patch j) onto the hemisphere placed on area G_i (patch i) is computed by shooting one ray from the center of patch i through each differential area ΔA_l of the hemisphere (see fig. 2b). If a ray intersects patch j, then its contribution δD_{ij} is added to D_{ij}. In case one ray has intersections with several patches, the contribution of the nearest patch is taken.

Shooting the entire hemisphere rays from an arbitrary patch is done in the following way. The set of hemisphere rays is created in advance, assuming an initial orientation along direction vector $(0\ 0\ 1)^T$. The direction vector (X Y Z) and delta form-factor assigned to rays are stored for each ray. Then the hemisphere is placed on the patch by rotating all hemisphere rays over the angle between the z-axis and the patch normal, using two Cordic rotation processors [2][13]. The Cordic is capable of rotating a 2 dimensional vector (figure 6). The rotation angles θ and ϕ are computed from $(0\ 0\ 1)^T$ and the patch normal $(nx\ ny\ nz)^T$ by another Cordic, which is not indicated in figure 6. $\phi = \arctan(nx^{-1} \cdot ny)$, $\theta = \arctan(nz^{-1}(nx \cdot \cos\phi + ny \cdot \sin\phi))$, and the rotated vector becomes:

$$(X_{\theta,\phi}, Y_{\phi}, Z_{\theta}) = Rot_{\theta,\phi}(XYZ).$$

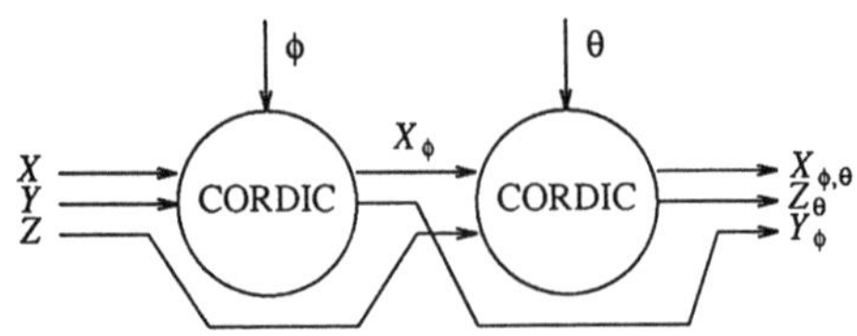

Figure 6. Rotation section

4. Adaptive Patch Sampling

In order to improve image quality, a patch can be subdivided into subpatches (or elements) in case it exhibits a high intensity gradient. In [4] this is done by using an adaptive patch subdivision method. This approach requires adjacency information, which is used to ensure the continuity of the intensity along the element edges in the rendering stage. Rather than subdividing a patch into smaller patches or elements, we compute the center-point of each of these elements, by which the number of patches remains constant. The center-points are obtained by using an "adaptive patch sampling" method. Before an outline of this method can be given, some background information regarding the substructuring technique is given first. For more details we refer to [4,14].

4.1 Substructuring

In the substructuring technique, element-to-patch form-factors instead of element-to-element form-factors are considered. The patch-to-patch form-factor is obtained by taking the area weighted average of the element-to-patch form-factors, that is:

$$F_{ij} = \frac{1}{A_i} \sum_{q=1}^{R} F_{iq,j} \Delta A_{iq} \tag{4.1}$$

where

$F_{iq,j}$ = the form-factor between element q (of patch i) and patch j.
F_{ij} = the form-factor between patch i and patch j.
R = the number of elements on patch i.

The coarse intensities (I_i) of the original patches are computed from the set of equations (2.6). From these intensities, the element intensities are obtained by using element-to-patch form-factors:

$$I_{iq} = F_{iq,j}.I_j \qquad (4.2)$$

where

I_{iq} = the intensity of element q of patch i.
I_j = the coarse intensity of patch j.

These element intensities accurately reflect the high intensity gradients on patches. The alternative patch sampling method is described in the next section.

4.2 Patch-Sampling

The idea behind the patch sampling method is based on the fact that actually only the center point of each element and the element-to-patch area ratio are required to compute form-factors. The elements referred to in this section are imaginary elements; their vertices are not relevant during the adaptive patch sampling process. The element center-points (or patch sample points) are defined in a global (x,y,z) coordinate system and their locations on the patch are kept up with local (u,v) coordinates. Both sets of coordinates are determined by using the binary subdivision technique. The ratio of the areas of a element and the patch, specifies the contribution of an element form-factor to the patch form-factor. In fact, this ratio can be interpreted as a weight-factor that can easily be approximated during the subdivision process. Let n be the number of subdivisions needed to obtain a particular imaginary element (q) and let the area of the patch i in the (u,v) coordinate system be normalized to 1, then the weight-factor, W_{iq}, that has to be attached to the corresponding sample point is 2^{-2n}. Hence, formula (4.1) can equivalently be written as:

$$F_{ij} = \sum_{q=1}^{R} W_{iq} F_{iq,j} \qquad (4.3)$$

Before starting the patch-sampling procedure, an initial guess must be provided for the number of sample points for each patch on which high intensity gradients are expected. From these points, form-factor computations are carried out assuming (temporarily) that the intensity of an area ΔA_{iq} surrounding sample point q of patch i is constant. For all patches and for all sample points, the form-factors $F_{iq,j}$ are determined. Applying (4.3), (2.6) and (4.2) yields one or more intensity values for each patch. These intensity values are obtained for an initial estimate of the possible gradient that might occur on a patch. To find out whether many samples are needed on a patch, a gradient test algorithm is invoked. This algorithm detects areas with high intensity gradient that have to be sampled finer in a next step. To get the new sample points, a patch is subdivided one step further. After this, the new form-factors for these points are computed and added to the element-to-patch form-factor matrix by removing the old row and inserting the four new rows. Naturally these rows provide more accurate element-to-patch form-factors. Again, applying (4.3), (2.6) and (4.2) will result in more intensity values per patch and hence will lead to a higher image quality. The whole procedure is adaptively repeated until the required criterion is met or the maximum subdivision depth is reached.

In case when there is an extreme high intensity gradient, taking more samples locally is not practical, since the intensity data per patch would increase drastically. A better solution here is to cut the patch along the sharp edge.

4.3 Gradient Test

To perform the gradient test, a 4-sample-point window is placed on the patch and a comparison procedure is activated to find out whether there is a high gradient or not. This procedure locally compares the differences among the intensity values against a prespecified margin. In case of a high intensity gradient, the whole patch is further subdivided to obtain a new set of center-points. If a window lies on a part of the patch with a high intensity gradient, then intensities for these new center-points on that part are computed via form-factor computations. On the other hand, if the window lies on a part of the patch that has no high intensity gradient, then intensities of the new center-points on these parts are obtained via bi-linear interpolation [5] of the intensities of the previous step. Although the subdivision of the whole patch is not necessary in case the gradient is local, it turns out that this procedure not only simplifies the gradient test algorithm, but also provides for a uniform intensity grid which is anyhow desired for the rendering process. In order to minimize storage requirements, one can also store, for each patch intensities hierarchically and perform the interpolations on the fly during the rendering stage.

When the intensity values for the new generated points are known, the gradient test algorithm is applied again. If the required criterion (margin) is not met, the whole procedure will be repeated. At termination, the adaptive patch sampling procedure provides a set of intensity values for each patch on which a high intensity gradient was expected. Finally a grid is placed on each patch such that each mesh of the grid embraces 4 sample points, and grid vertex intensities are derived from the sample point intensities by interpolation. Only vertex intensities are stored; this reduces overhead. Moreover, in the rendering stage interpolation for pixel intensity computation, is now done between the vertices; this is simpler than between sample-points.

5. Rendering

The preprocess provides an intensity grid for each patch. The resolution of the grid varies depending on the intensity gradients on patches.

The postprocess computes the actual pixel intensities for red, green and blue by tracing one ray through each pixel of the displaying screen. If the ray hits a patch, two different (u,v) coordinates are reported to the host. The first (u,v) pair is needed to indicate in which mesh an intersection point is located, and the second pair is needed to determine the actual intersection point within such a mesh. Then, to compute the diffuse intensity component for a screen pixel, we can interpolate between intensities of the corresponding mesh points by using both (u,v) coordinates. If the pixel ray hits a specular patch, then besides the interpolated diffuse intensity component, if any, the specular intensity component is computed by placing a frustum on the patch. If a frustum ray hits a specular patch, a frustum is placed again, and so on. The number of recursive steps can be controlled in two ways. One is by limiting the depth of recursion adaptively according to the amount of the contributed intensity. The other is by reducing the frustum size for successive bounces.

Notice that shooting rays trough the displaying screen is similar to shooting rays from the hemisphere and the computation of specular intensity components corresponds to the computation of indirect form-factors. This means that the same computer architecture can be used for both pre- and postprocess.

6. Hardware

At this stage it is not yet clear which tasks of the entire image rendering algorithm are suitable to be implemented in software or in hardware. As far as the hardware is concerned, we therefore restrict ourselves to the heart of the system which is the computation of form-factors and the solution of a set of equations. All other computations such as bilinear interpolation and adaptive patch-sampling will be assumed to be performed by the host.

Let R be the number of hemisphere rays and P the number of patches within the environment. For a purely diffuse environment a number of $P \cdot R$ rays will be rotated and shot. If space subdivision is not used, all patches must pass each ray, so the number of intersection computations becomes $P^2 \cdot R$, resulting into $P \cdot R$ intersections. The time needed to compute the form-factors is therefore proportional to $P^2 \cdot R$, which makes the intersection computation the most time consuming operation of the algorithm. Specular patches will cause an even greater number of rotation operations and intersection computations.

The rotation operations are performed using a special high-speed Cordic rotation processors [13]. Our goal therefore is to design an architecture built around several intersection computation blocks.

6.1 Intersection Computation

The hemisphere is rotated and placed on each patch. The rays that are shot have the patch center as their origin. Each ray can accordingly be defined as:

$$\vec{r} = \vec{O} + t \cdot \vec{a} = \begin{pmatrix} x_0 \\ y_0 \\ z_0 \end{pmatrix} + t \cdot \begin{pmatrix} a_x \\ a_y \\ a_z \end{pmatrix}, \text{ with } ||\vec{a}|| = 1.$$

In view of hardware implementation we chose an iterative bounding box based algorithm for the intersection computation, as follows.

REPEAT UNTIL STACK IS EMPTY OR ACCURACY CRITERIA ARE MET:
- POP PATCH FROM STACK
- SUBDIVIDE PATCH INTO FOUR SUBPATCHES
- FOR EACH SUBPATCH DO

BEGIN COMPUTE BOUNDING BOX :
$MAX_{X,Y,Z}$(CONTROLPOINTS)= $X_{MAX}, Y_{MAX}, Z_{MAX}$
$MIN_{X,Y,Z}$(CONTROLPOINTS)= $X_{MIN}, Y_{MIN}, Z_{MIN}$
COMPUTE SIX INTERSECTIONS OF BOUNDING BOX WITH RAY:
$t_{XMAX}, t_{YMAX}, t_{ZMAX}, t_{XMIN}, t_{YMIN}, t_{ZMIN}$; where $t_{XMAX} = a_x^{-1} \cdot (X_{MAX} - x_0)$
IF MIN($t_{XMAX}, t_{YMAX}, t_{ZMAX}$)≥ MAX($t_{XMIN}, t_{YMIN}, t_{ZMIN}$) THEN
hit:=true, PLACE SUBPATCH ON STACK, UPDATE (u,v) COORDINATES
END.

The advantages of this method are:

1. At termination, the algorithm produces the parametric (u,v) coordinates of the intersection point directly.
2. The t-value, that is the distance between the origin of the ray and the intersection point is automatically produced as well as the intersection point itself. The t-value is used for distance comparison, the intersection point for multiple reflection/transmission.
3. The accuracy can be varied by simply adjusting the maximum subdivision level and the minimum box dimension.

The system architecture used for the intersection computation is shown in figure 7, which is similar to the architecture for ray-tracing bicubic patches in Bezier form as proposed by Pulleyblank in [17]. See also section 7.

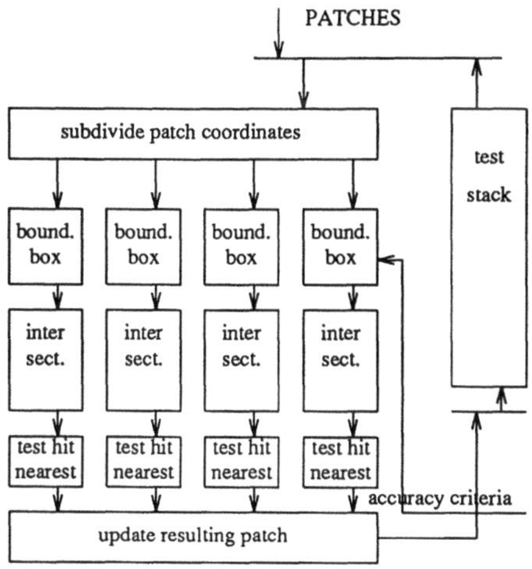

Figure 7. Intersection computation

6.2 Multiple Reflectance

As stated in section 3, due to multiple reflection and/or transmission each primary (hemisphere or pixel plane) ray can possibly generate a tree of frustum-rays to be shot. Such a tree could possibly become too large, which makes shooting all rays almost impossible. This problem is naturally solved using a backtracking procedure. The frustum axes, that is the reflected or transmitted rays that have the largest contribution of all frustum rays, form the tree's skeleton. The skeleton is first composed by only considering the axis of the reflected and/or transmitted frustums. When the contribution of an next axis has become no longer significant after several bounces, the skeleton stops. The last frustum axis is taken from the skeleton and its frustum rays are then shot. It may again be possible for one of these rays to be stored as another frustum axis. If all rays within a frustum are shot, then the frustum that is one level higher in the tree is shot, and so on. This backtracking procedure is carried out until all frustums have been shot for one primary ray. The skeleton must be stored together with all necessary information that is required to generate the frustum rays. The following values have to be stored:

— the frustum axis' origin and direction vector (or direction angles θ and ϕ),

— the frustum size and the number of rays within the frustum that already have been shot

6.3 Functional Block Diagram

A preliminary functional block diagram for the architecture is illustrated in figure 8. The intersection computation block, shown in figure 7, computes the intersection point as described in section 6.1 and passes the required information, such as the number of the intersected patch with its physical description, the intersection point and the contribution (delta form-factor) of the ray. In the k_d processor the direct contributions to the total form-factor are computed. The indirect contributions, via specular reflection and transmission, to the total form-factor are computed in the k_t and k_r processors respectively. If the contribution of the skeleton ray is greater than the threshold value Δ, the transmitted and/or reflected axis ray is computed, pushed on the stack with all necessary information (see sub section 6.2) and immediately shot again. When the contribution of the specular ray(s) is no longer significant, the skeleton is ended. The two circles labeled Cordic provides for the steering of the primary and secondary rays. The whole procedure is supervised by the block labeled control.

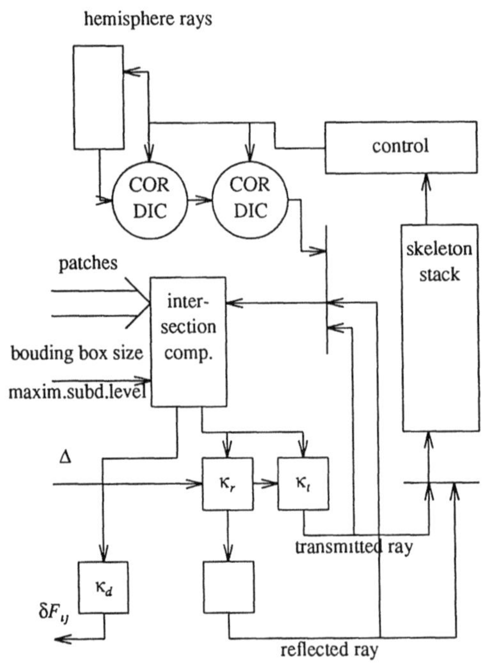

Figure 8. Functional block diagram

6.4 Solving the Radiosity Equations

A less time consuming stage of the algorithm is the solving of the radiosity equations (2.6), using an iterative Gauss-Seidel method. A hardware definition for this stage has already been proposed in [2]. The architecture presented there is built around ASIC processing elements (PEs), which are (pipelined) multiplier/accumulator-processors.

6.5 Hardware Considerations

If the threshold value Δ is kept independent of the frustum size, the error made by the selection of rays to be shot could become very large if the number of rays within one frustum is large. The threshold value Δ should therefore take on a value depending on the number of rays within a frustum. The complexity of the architecture depends on the required accuracy for the intensities of the screen pixel. The screen resolution determines the number of bits needed to express the pixel intensities and the form-factors, which also indirectly determines the number of rays that must be shot from one hemisphere. A thorough investigation in this area is most necessary. Among several further considerations, the most important ones are the following:

— The computation of the reflected and transmitted rays (see figure 8) could be done on board or externally using an extra Cordic-processor.

— The stack for saving the skeleton could become too large to stay on board.

— The subdivision algorithm automatically produces the (u,v) coordinates of the intersection points, which can be used for interpolation when using quadrangular patches. When using different, e.g. triangular patches, another interpolation procedure has to be used.

7. Bezier Patches

A 3D Bezier curve is defined by 16 control points, of which only the 4 corner points actually lie on the surface. A surface representation can be found by subdivision using the control points [17]. After one subdivision step the resulting points define four Bezier sub patches, each determined again by 16 points. The subdivision is repeated until the sub patches may be considered having a constant intensity. In the radiosity method the intensity of each patch is only computed for the center-point of this patch. Without loss of accuracy the leaf patch takes on the intensity value computed for its center-point. In order to compute the form-factors between two arbitrary Bezier patches, we used only 16 center-points and their 16 normals. The hemisphere is placed on the patches centered around each of the 16 points. Let u and v be the parameters along the Bezier patch, we obtain after 3 subdivisions 16 points lying on the surface with (u,v) coordinates $u = 2(n+1/2)/8$, $v = 2(m+1/2)/8$, with $0 \leq m, n \leq 3$. The normal to the surface in such a point is found from the four corner points obtained after 2 subdivisions, by computing the cross product of the two estimated tangent-vectors.

The ray-patch intersection computation is carried out by an iterative bounding box procedure as described in section 6.1. The advantage of this procedure is that the (u,v) coordinates of the intersection point are directly available. The (u,v) coordinates are used to determinate which of the 16 subpatches is hit. The u,v coordinates are also used during the rendering stage, where again a bilinear interpolation procedure is applied. The rendering of shadows across the surface can be accomplished by adaptive patch sampling, similarly as for the polygonal patches (see section 4).

8. Statistics and Simulations

With a software version of the algorithm some simulations have been carried out for various values of the parameters, such as the number of hemisphere rays, number of hemisphere circles and the radiosity of patches. In figure 9a is shown how two radiosity values for RED (b_R), for two arbitrary patches, converge as the number of hemisphere circles R increases. In figure 9b is shown how the total number of hemisphere rays depends on the number of hemisphere circles used. The time needed to compute the form-factors is proportional to the number of hemisphere rays (section 6).

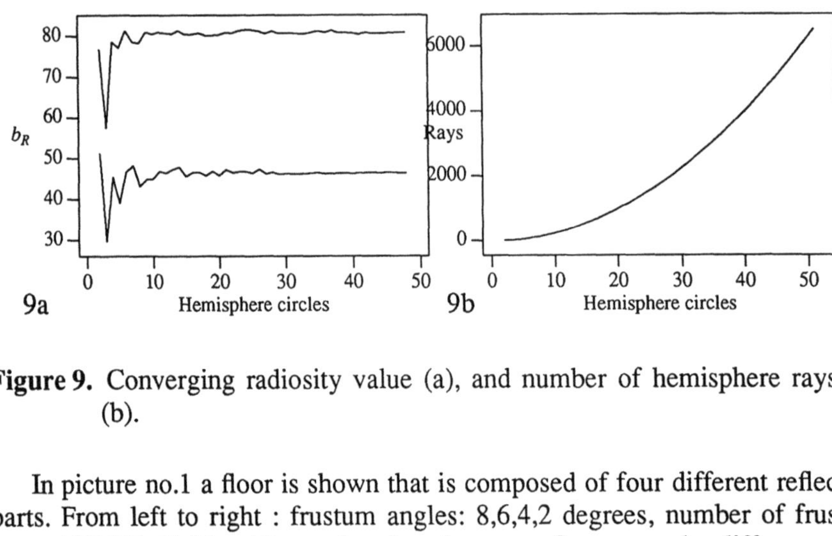

Figure 9. Converging radiosity value (a), and number of hemisphere rays (b).

In picture no.1 a floor is shown that is composed of four different reflecting parts. From left to right : frustum angles: 8,6,4,2 degrees, number of frustum rays: 195,121,61,19. All patches in picture no.2 are purely diffuse and/or perfect mirror reflecting. The two-pass approach proves to be an efficient method to integrate diffuse and specular reflectance behavior. The computation of indirect form-factors accounts for the extra amount of light on the floor directly in front of the mirror. A Bezier teapot consisting of 39 Bezier patches is shown in picture no.3. The number of form-factors determined for this picture is $(39*16)^2$. An approximation of each Bezier patch by 50 triangular

Picture 1

Picture 3

Picture 2

Picture 4

patches would have required $(39*50)^2$ form-factors to be computed, which implies 13 times as much computation time and 3 times as much memory demand. Picture no.4 enlights the patch sampling method. The maximum subdivision depth has been chosen 4. The number of form-factors, that is the number of sample points computed for the floor equals 28.

9. Conclusions

In this paper we have presented a hardware algorithm for the synthesis of realistic 3D images on an engineering workstation.

Both form-factor computation and image rendering (including intensity computation on specular surfaces) are based on ray-tracing. The computation of patch to hemicube projection has been replaced by an estimation of patch to hemisphere projection, which is particular suitable for ASIC implementation. Moreover, this approach allows a simple selection of a delta form-factor representing an amount of energy within a small solid angle originating directly from a diffuse patch, or indirectly via specular reflective patches.

Shadows along surfaces are computed by applying "adaptive patch sampling", indicating that the number of points on patches grows but not the number of patches, thereby saving local and global memory.

A set of rays defining the hemisphere are stored in advance. For each diffuse patch they are rotated according to the patch orientation and cast independently. Cordic arithmetic is used to perform the rotation operations. A bounding box subdivision procedure is used to compute the ray-patch intersection point. The patches may have polygonal as well as Bezier forms.

Since the form-factor computation in the preprocess and the image rendering in the postprocess are solely ray-tracing based, a single SIMD/wavefront-type architecture is naturally used to implement both processes. This architecture, when built on ray-rotating and ray-tracing pipelined processors allows for a very high throughput VLSI implementation of the generalized two-pass procedure.

As mentioned before, a set of accuracy criteria must be decided upon before a final architecture can be designed. Also the possibility and consequences of including non uniform rational Bezier curves (NURB)s has to be investigated.

The (u,v) coordinates play an important role during several stages of the algorithm. The subdivision hardware that was proposed in this paper is very suitable to generate these coordinates, but it is not suitable for pipelining the operations. This problem can be overcome by modifying the subdivision strategy.

Since the size of the form-factor matrix can not grow larger than the available storage capacity, a restriction to the number of patches within environments may be necessary. This restriction can be removed by using a progressive refinement technique. This has already been presented in [6] for diffuse surfaces. The algorithm presented in our paper can be easily adjusted in that way. The final architecture of the generalized two-pass 3D image synthesis engine will also allow for an extended version of the progressive refinement method given in [6].

10. Acknowledgments

The authors wish to thank Professor F.W. Jansen of the department of Computer Sciences at the Delft University of Technology (The Netherlands), for many valuable discussions. They also whish to thank Erik Platzbecker for his contribution to the software.

11. References

1. Blinn, J.F., "Models of Light Reflection for Computer Synthesized pictures," Proceedings of SIGGRAPH'77, In Computer Graphics, Vol. 11, No. 2, 1977, pp. 192-198.
2. Bu J.and E.F.Deprettere, "A VLSI System Architecture for High-Speed Radiative Transfer 3D Image Synthesis", The Visual Computer, Vol. 5 No. 3 June 1989.
3. Bu J. and E.F.Deprettere, "A Parallel VLSI Algorithm for Fast Sparse Matrix Solution by Gauss-Seidel Iteration". Proc. ISCAS '87, Vol. 3, pp.1052-1055, 1987.
4. Cohen, M.F, Greenberg, D.P, Immel, D.S, Boch, P.J, "An Efficient Radiosity Approach for Realistic Image Synthesis", IEEE Computer Graphics & Applications Vol.6 No.2 March 1986 pp.26-35.
5. Cohen, M.F, Greenberg, D.P, "The Hemi-Cube, A Radiosity Solution for Complex Environments", ACM Proceedings of SIGGGRAPH '85 Vol.19 No.3, July 1985 pp.31-40.
6. Cohen, M.F, Chen, S.E, Wallace, J.R, Greenberg, D.P, "A Progressive Refinement Approach to Fast Radiosity Image Generation", Computer Graphics, Vol. 22, No. 4, August 1988, pp. 75-84.
7. Cook, R.L, Porter, T, Carpenter, L, "Distributed Ray-Tracing", Computer Graphics 18,3,pp.137-146, 1984.
8. Foley, J.D. and Dam, van A, "Fundamentals of Computer Graphics". Addison-Wesley Publishing Co., 1882.

9. Goral, Cindy M., Kenneth E. Torrance, Donald P. Greenberg, Bennet Battaile, "Modelling the Interaction of Light Between Diffuse Surfaces", Proceedings of SIGGRAPH'84, In Computer Graphics, Vol. 18, No. 3, July 1984, pp. 213-222.
10. Hall, R.A. and Greenberg, D.P., "A Testbed for Realistic Image Synthesis", IEEE Computer Graphics and Applications, Vol. 3, No. 10, Nov. 1983, pp. 10-20.
11. Immel, David S., Micheal F. Cohen, Donald P. Greenberg, "A Radiosity Method for Non-diffuse Environments", Proceedings of SIGGRAPH'86, In Computer Graphics, Vol. 20, No. 4, Aug. 1986, pp. 133-142.
12. Kajiya, J.T., "The rendering Equation", Computer Graphics, Proceedings Siggraph 86, Vol.20, No.4, August 1986, pp. 143-150.
13. Lange de A.A.J., Hoeven van der A.J., Deprettere E.F., Bu J., "An Optimal Floating-point Pipeline CMOS CORDIC Processor", Proc. ISCAS '89, Vol. 3, pp. 2043-2048, 1988.
14. McGuire W. Gallagher R.H. "Matrix Structural Analysis", John Wiley & Sons, New York, 1979.
15. Min-Zhi Shao, Qun-Sheng Peng, You-Dong Liang, "A New Radiosity Approach by Procedural Refinements for Realistic Image Synthesis", Computer Graphics, Vol. 22, No. 4, August 1988, pp. 93-101.
16. Phong, Bui Tuong, "Illumination for Computer Generated Pictures," Communications of the ACM, Vol. 18, No. 6, June 1975, pp. 311-317.
17. Pulleyblank, R.W, "The Feasibility of a VLSI Chip for Ray Tracing Bicubic Patches", IEEE CG&A, march 1987 pp.33-44.
18. Siegel R. and Howell J.R, "Thermal Radiation Heat Transfer", Hemisphere Publishing Corp., Washington DC.,1981.
19. Sparrow, E.M. and Cess, R.D. "Radiation Heat Transfer", Hemisphere Publishing Corparation Washington DC., 1978.
20. Wallace, J.R, Cohen, M.F, Greenberg, D.P, "A Two Pass Solution to the Rendering Equation - A Synthesis of Ray-Tracing and Radiosity Methods", ACM Proceedings of SIGGRAPH'87 Vol. 21, No. 4, July 1987, pp. 311-320.

The HERO Algorithm for Ray-Tracing Octrees

Mark Agate, Richard L. Grimsdale and Paul F. Lister

VLSI & Graphics Research Group
School of Engineering, University of Sussex
Brighton BN1 9QT, UK

Abstract

An algorithm is presented for rapid traversal of octree data structures, in order to enhance the speed of ray tracing for scenes of high complexity. At each level of the octree, the algorithm generates the addresses of child voxels in the order they are penetrated by the ray. This requires only a few arithmetic operations and simple logical operations. A depth-first search of the tree is used to yield the first terminal voxel hit by the ray, thus hidden objects are not processed. The algorithm is designed specifically for implementation as HERO: A Hardware Enhancer for Ray-tracing Octrees.

1. Introduction

Computation of ray/object intersections is widely recognised as being the major bottleneck in ray tracing. Even when object descriptions are made as simple as possible, ray tracing a scene of even moderate complexity can be an extremely slow operation, due to the nesting of two large loops (one for all rays, the other for all objects). Plunkett and Bailey [1] have applied array processing techniques to enhance the performance of a ray tracing system, but even with the power of a supercomputer, the performance falls short of real-time. Attempts to speed up ray tracing significantly must therefore be based on algorithmic developments, exploiting either image coherence to reduce the complexity of the outer loop, or object space coherence to reduce the number of intersections required in the inner loop.

To exploit image coherence, it is necessary to use a beam tracing approach so that a whole area of the image can be processed at a time. This is done by Heckbert and Hanrahan [2], who cast a single rectangular pyramid for the entire screen, recursively subdividing this into smaller pyramidal beams wherever the image is complex. Whilst this can be expected to yield a considerable speedup over conventional ray tracing if large areas of the image are homogeneous, a characteristic of ray traced images is the lack of image coherence which results from rendering of surface characteristics and illumination effects. Pyramidal beam intersections, though advantageous for antialiasing, are considerably more complex to compute than point sampling with rays, and the extra complexity of recursive beam subdivision makes this approach undesirable for direct implementation in hardware.

Object coherence is partly exploited by Whitted [3], who uses bounding volumes to reduce the complexity of ray/object intersection calculations, though the size of the inner loop is still heavily dependent on the number of objects, since every spherical bounding volume must still be processed for every ray. Kay and Kajiya [4] use hierarchical object bounding, whereby objects made up of several smaller objects are bounded by parallelopipeds which are stored in a tree structure. A ray which fails to penetrate the outermost bounding volume of an object need not be intersected with any volumes or objects inside it, and the complexity of the inner loop is thereby reduced. A similar approach is used by Dippe and Swensen [5], though the shapes of the bounding volumes themselves are varied, resulting in more complex ray/volume intersection calculations.

In a similar way to that in which Bresenham's algorithm generates the locations of 2-D pixels which are intersected by a line, Amanatides and Woo [6] have developed an incremental algorithm which returns the 3-D voxels hit by a ray. As well as eliminating ray intersection calculations with objects which are far from the line of the ray, an incremental voxel tracer can terminate as soon as the first intersection with an object is found, without having to intersect the ray with other objects and subsequently find the nearest. Occluded objects therefore remain unprocessed.

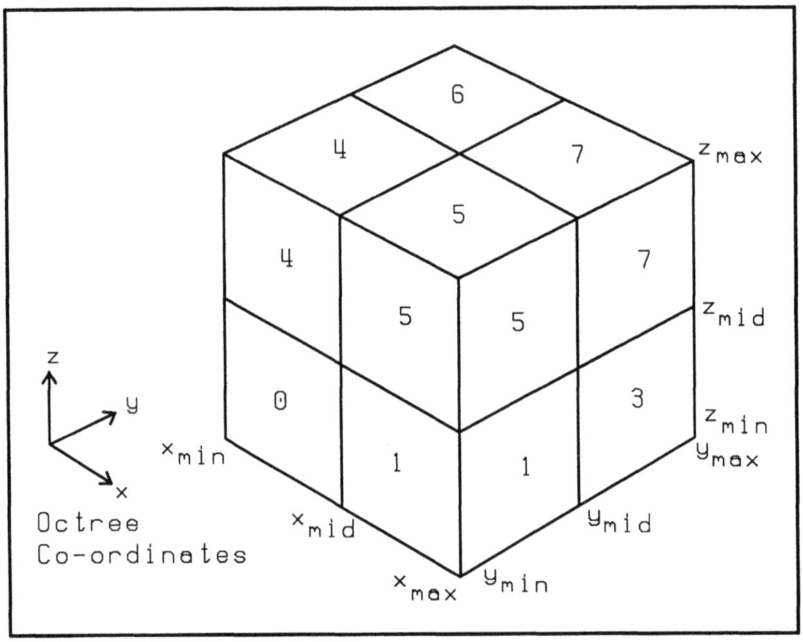

Figure 1 Octree Definition

Further economy in ray/object intersection calculations can be afforded if it is the object space, rather than the object descriptions, which is subdivided hierarchically. The most common technique for space subdivision is the generation of an octree, as illustrated in fig. 1. Voxels are stored in a tree structure, each node having between zero and eight children. A voxel containing only a simple object or surface has no children, but if its contents are more complex than this, it is subdivided into eight child voxels as shown. Any child voxels which are empty can be eliminated, those which are simple form leaf nodes, and others are subdivided further. Note that the voxels in an octree may be of many different sizes. When such a structure is ray traced, voxel contents can be searched in the order that the voxels are hit by the ray, thus leading straight to the first object hit. An additional bonus with octrees is that large areas of empty space are rapidly traversed by the ray.

Glassner [7] uses a scheme in which the child voxels of a node are numbered from 1 to 8. The ray is traced through the octree by moving along the line of the ray a certain distance, then determining which voxel the resulting point is inside. Although a considerable speedup is obtained over conventional ray tracing, this scheme is not ideal for hardware implementation, due to the voxel numbering system adopted. It is shown in the next section that simple logic operations may be employed for voxel address generation if an octal system is used instead, with nodes labelled from 0 to 7.

Fujimoto et al [8] trace rays through voxels using two incremental line-drawing type processors in a similar way to Amanatides and Woo, but within an octree data structure. It is observed that the image rendering time is "Virtually independent of the number of objects in the scene", and that for very large numbers of objects, this technique of ray tracing actually becomes faster than projective methods. Peng et al [9] use an approach to voxel traversal which appears to be essentially heuristic, in that the ray is assumed to pass to the neighbouring voxel in the direction of the maximum ray direction co-ordinate; if this is not found to be the case, the second most likely neighbour is tested, and if this fails, the least likely neighbour is hit. All these techniques undoubtedly provide considerable speed enhancements. However, the algorithm presented here removes the need to test a ray against a voxel, since the addresses of the child nodes are generated in the order they are hit. This entails only a few arithmetic operations for each parent voxel, and simple logic operations for address generation. These operations are ideally suited to hardware implementation, thus HERO (a Hardware Enhancer for Ray-tracing Octrees) has been conceived.

2. Development of the HERO Algorithm

An efficient way to represent a convex object with planar faces is to define it as the intersection of half-spaces. (A half-space is the volume of space to one side of an infinite plane). The cube in fig. 2 is defined by six bounding planes, each with a unit normal vector $\underline{c}_i$ pointing away from the interior of the object, and position constant d_i, so that all points $\underline{P}$ on a particular bounding plane satisfy the equation:

$$\underline{c}_i . \underline{P} + d_i = 0$$

Thus each face of a solid modelled in this way is geometrically defined by just four numeric values.

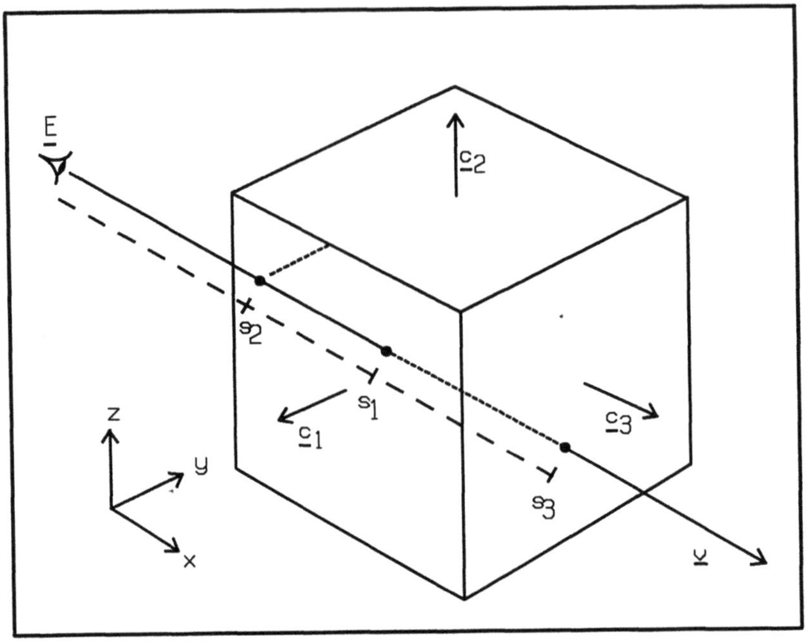

Figure 2 Cube Defined by Half-Spaces

All points $\underline{P}$ along the line of a ray can be defined by the equation:

$$\underline{P} = \underline{E} + s * \underline{v}$$

where $\underline{E}$ is the ray origin position, $\underline{v}$ is the direction vector and s is a positive scalar variable representing the distance along the ray. The ray hits an object defined in the above way only if there exists a range of points on the line of the ray which are simultaneously to the inside of all the bounding planes for that object. For any one bounding plane, the distance parameter s_i to the point at which the ray hits the plane ($\underline{c}_i . \underline{P} + d_i = 0$) can be found from:

$$s_i = -(\underline{c}_i . \underline{E} + d_i)/(\underline{c}_i . \underline{v})$$

If the angle between the ray and the surface normal is obtuse, points on the ray at distances greater than s_i from the ray origin will be inside the half-space defined by the bounding plane. Conversely, if this angle is acute, only points with $s < s_i$ will be inside the half-space. Thus for each bounding plane, a positive value of the dot product $\underline{c}_i . \underline{v}$ indicates that s_i is an upper limit on the required range of distances, while a negative value indicates that s_i is a lower limit. To determine if the ray hits the object, all s_i values are computed and categorised as upper or lower limits. The maximum lower limit (s_{lmax}) and the minimum upper limit (s_{umin}) are then found; if $s_{lmax} < s_{umin}$, there exists a range of distances for which points on the ray are internal to all the half-spaces. If both limits are positive, then a forward-travelling ray hits the object at the point:

$$\underline{P} = \underline{E} + s_{lmax} * \underline{v}$$

In the example shown in fig. 2, s_{lmax} is equivalent to s_1, and s_{umin} is s_3. From $s_1 < s_3$, it may be deduced that the ray penetrates the cube, and does so at a distance s_1 from the ray origin.

It is important to consider the effects of numeric overflow in the calculation of distance values; this can occur if the ray is parallel or nearly parallel to the bounding plane under consideration, causing the value of $\underline{c}_i \cdot \underline{v}$ to become very small. In such cases, even if the value of s_i should have a magnitude of infinity, it is sufficient to represent this by the maximum numeric magnitude allowed, as long as the sign of the result is correctly evaluated. Fig. 3 demonstrates two such situations. In both cases, $\underline{c}_i \cdot \underline{v}$ is zero, since the rays lie parallel to the bounding plane. Assuming that this zero value is represented as positive, both values of s_i will be interpreted as upper limits on s. In the first case, the value of $\underline{c}_i \cdot \underline{E}_1 + d_i$ is positive (since $\underline{c}_i$ points towards $\underline{E}_1$), hence the computed value for s_i will be minus infinity. Since this is an upper limit on s, it is clear that the ray does not penetrate the half-space defined by the bounding plane along its entire length. In the second case, $\underline{c}_i \cdot \underline{E}_2 + d_i$ is negative (as $\underline{c}_i$ points away from $\underline{E}_2$), thus a value of plus infinity is returned for s_i, indicating that the ray is inside the defined half-space along its entire length. Conversely, if $\underline{c}_i \cdot \underline{v}$ were in fact a small negative value, the sign of s_i would be inverted in each of the above cases, but these would be interpreted as lower limits on s, hence the system is well-behaved.

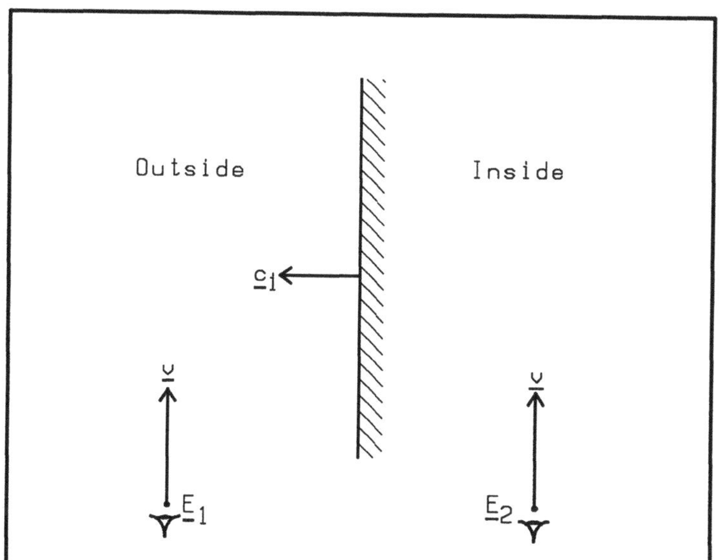

Figure 3 Rays Parallel to Bounding Planes

The root node of an octree may be defined in the same way as the cube of fig. 2. If the octree and ray are both in a co-ordinate system which has axes parallel to the edges of the octree root node, as in fig. 1, then all unit normal vectors $\underline{c}_i$ will have two co-ordinates equal to 0 and one equal to plus or minus unity. The dot product operation in computation of s distances can therefore be reduced to a simple co-ordinate selection and possible sign inversion.

The bounding planes for each node of the octree can be labelled x_{min}, x_{max}, y_{min}, y_{max}, z_{min} and z_{max}, as in fig. 1. The planes $x = x_{mid}$, $y = y_{mid}$ and $z = z_{mid}$ define the new bounding planes for the child voxels of the parent node. Each child voxel has a numeric identifier between 0 and 7 as shown, which can be represented by a 3-bit binary word. If the three bits are considered to correspond to the co-ordinates z, y, x in order, then a '0' digit in the child's identifier indicates that it shares the parent's min boundary for that co-ordinate, while a '1' indicates that it shares its parent's max. For example, child node 3 (mask = 011) shares its parent's z_{min}, y_{max} and x_{max} boundaries. The child node's z_{max} value will be its parent's z_{mid}, its y_{min} will be its parent's y_{mid}, and its x_{min} is its parent's x_{mid}.

If a ray is found to penetrate the root node of the octree, it is subsequently necessary to determine which child voxels are hit by the ray. A simplistic approach would be to test each of the non-empty child voxels for penetration by the ray using the half-space method, then sort them by their $s_{l\,max}$ values to determine the order in which they are hit. It should be noted that the ray need only be intersected with the parent's x_{mid}, y_{mid} and z_{mid} planes once; the 3-bit child node identifiers can then be used to determine which distance values apply to each child.

A further simplification may be afforded by noting that the order in which child nodes are hit can be found by testing the children in the correct order. The order in which they should be tested depends solely on the octant of space into which the ray points. For example, if the x_v, y_v and z_v components of the ray direction vector $\underline{v}$ are all positive, then the child nodes can be tested in the order 01234567. If (say) nodes 5, 7 and 1 are found to be hit, these must be penetrated in the order 1, 5, 7. As another example, suppose x_v and y_v are positive, but z_v is negative; the order in which child nodes should be tested is then 45670123. A technique for generation of these orderings from the ray direction vector is as follows.

For the ray direction vector $\underline{v}$, generate a 3-bit 'VMASK', with individual bits corresponding to sign(z_v), sign(y_v) and sign(x_v); '0' for positive values and '1' for negative. The order in which child nodes should be tested can be found by running a counter from 0 to 7, and returning the value of COUNTER XOR VMASK. For example, a ray with negative x_v but positive y_v and z_v would have VMASK = 001, giving rise to the child node ordering 10325476.

The next stage in simplifying the algorithm is to notice that, for any parent node intersected by the ray, between 1 and 4 of its child nodes will be intersected. Additionally, only certain combinations of child voxels can be hit. For the example shown in fig. 4, with VMASK = 000, once it is found that child node 1 is intersected, it is known that nodes 0, 2, 4 and 6 cannot be hit. Logical derivation of the order of nodes hit can in fact be achieved in the following way.

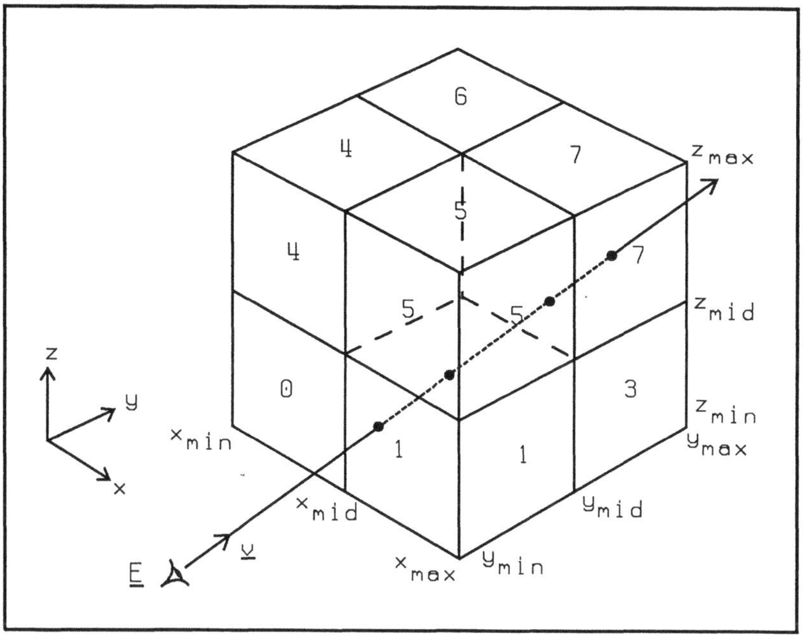

Figure 4 HERO Example 1

For the simple case VMASK = 000, let the identifier of the first child hit be CHILDMASK. The individual bits of CHILDMASK are set according to the following conditions ('0' for false, '1' for true):

$$\text{CHILDMASK} = s_{zmid} < s_{lmax},\ s_{ymid} < s_{lmax},\ s_{xmid} < s_{lmax}$$

where all s values are computed for the parent node. It can be seen that the identifiers of subsequent children hit (if any) can be found by setting the as yet unset bits of CHILDMASK. For example, if node 1 is the first child hit (CHILDMASK = 001), as in fig. 4, then the only other children which can be intersected have CHILDMASK equal to 011, 101 or 111 (i.e. nodes 3, 5 or 7). Similarly if node 5 were the first child hit, only node 7 could subsequently be hit. The order in which these bits should be set is found as follows:

1. For the parent node, compute the distances along the ray to the mid-plane intersections, i.e. s_{xmid}, s_{ymid} and s_{zmid}.

2. Sort these s values into ascending order. Enter the corresponding masks - 001 for s_{xmid}, 010 for s_{ymid} and 100 for s_{zmid} - into a list which indicates the ascending order of s values. In the example shown in fig. 4, where $s_{xmid} < s_{zmid} < s_{ymid}$, this list will be as follows:

   MASKLIST [1] = 001
   MASKLIST [2] = 100
   MASKLIST [3] = 010

3. Take each element of MASKLIST in turn, and if the corresponding bit of CHILDMASK is not already set, that element should be ORed into CHILDMASK to produce the identifier of the next child hit. Thus for fig. 4, the initial value of CHILDMASK is 001, and subsequent identifiers 101 and 111 are produced, by ORing in elements 2 and 3 of MASKLIST, respectively.

Two further considerations are necessary for generalisation of the algorithm. The first is that the generation of new child identifiers should stop when the last child voxel is hit. In a similar way to that in which CHILDMASK is generated, the mask for the last child hit can be generated by:

$$\text{LASTMASK} = s_{zmid} < s_{umin},\ s_{ymid} < s_{umin},\ s_{xmid} < s_{umin}$$

The termination condition can therefore be easily set. Secondly, the analysis so far has concentrated on child voxels hit by a ray in the octant with VMASK = 000. For rays in other octants, each new CHILDMASK generated should be XORed with VMASK to produce the actual identifier of the next child hit. In effect, CHILDMASK replaces the counter which was used to determine the order in which child voxels should be tested. The XOR operation implements a transformation of CHILDMASK into the correct octant of space, according to the direction of the ray.

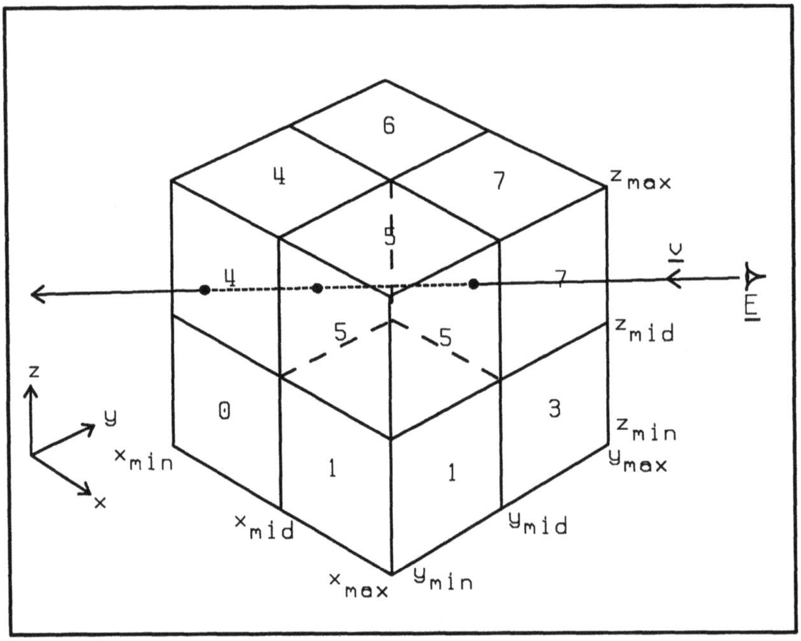

Figure 5 HERO Example 2

A second example, shown in fig. 5, has VMASK equal to 111. The distance s_{lmax} is equivalent to s_{xmax}, and s_{umin} is equivalent to s_{ymin}. The value of CHILDMASK is initially 010, and LASTMASK is 011. Since $s_{ymid} < s_{xmid} < s_{zmid}$, MASKLIST is as follows:

MASKLIST[1] = 010
MASKLIST[2] = 001
MASKLIST[3] = 100

The first child hit therefore has an identifier equal to the initial value of CHILDMASK XOR VMASK, i.e. 010 XOR 111, which is 5. A logical OR of MASKLIST[1] into CHILDMASK does not create a new value, but when MASKLIST[2] is ORed in, CHILDMASK becomes 011, which is equal to LASTMASK. Hence the second and final child hit has identifier 011 XOR 111, giving the correct value of 4.

3. Summary of the Algorithm

For the ray $\underline{E} + s * \underline{v}$, the following 3-bit mask is generated:

$$\text{VMASK} := \text{Sign}(z_v), \ \text{Sign}(y_v), \ \text{Sign}(x_v)$$

For the root node of the octree, the distance parameter s for each of the six bounding planes is found:

$$s_{xmin} := (x_{min} - x_E)/x_v \quad ; \text{etc.}$$

Each s value is categorised as either an upper limit or lower limit on the range of s for points inside the root node. A '0' in the x-bit position of VMASK implies that s_{xmin} is a lower limit and s_{xmax} an upper limit, while a '1' implies the opposite. The s values for the y and z planes are similarly categorised. The maximum of the lower limits (s_{lmax}) and the minimum of the upper limits (s_{umin}) are found. If $s_{lmax} < s_{umin}$, the root node is penetrated by the ray, otherwise it is not.

For any parent node intersected by the ray, the distance values for intersection of the ray with each of the node's mid-planes are calculated, i.e. s_{xmid}, s_{ymid} and s_{zmid}. These are sorted into ascending order, and the masks 001 for s_{xmid}, 010 for s_{ymid} and 100 for s_{zmid} are entered into MASKLIST[1..3] in ascending order of their corresponding s values. Two further masks are also generated:-

$$\text{CHILDMASK} := (s_{zmid} < s_{lmax}), \quad (s_{ymid} < s_{lmax}), \quad (s_{xmid} < s_{lmax}) \ ;$$
$$\text{LASTMASK} := (s_{zmid} < s_{umin}), \quad (s_{ymid} < s_{umin}), \quad (s_{xmid} < s_{umin}) \ ;$$

The following operations then generate the identifiers of the child nodes in the order they are hit:

```
i := 1 (* index into MASKLIST *)
LOOP
   SearchChild(CHILDMASK XOR VMASK);
      (* check next intersected child *)
   IF CHILDMASK = LASTMASK THEN EXIT;
      (* all intersected children searched *)
   ELSE
      WHILE (MASKLIST[i] AND CHILDMASK) <> 0 DO
         (* find next valid masklist element *)
         i := i + 1;
      END; (* WHILE *)
      (* now generate new CHILDMASK *)
      CHILDMASK := CHILDMASK OR MASKLIST[i];
   END; (*ELSE *)
END; (*LOOP *)
```

While the above demonstrates how the correct child node identifiers can be generated, it is also necessary to associate the proper values for s_{lmax} and s_{umin} with each intersected child. For any parent node, the first child hit shares its parent's s_{lmax}, and the last child hit shares its parent's s_{umin}. If more than one child is hit, the s_{umin} value for all but the last is equal to either s_{xmid}, s_{ymid} or s_{zmid}. The correct value is selected according to the entry in MASKLIST used to generate the subsequent child hit. For example, if the CHILDMASK of the next child hit is generated by ORing in the mask 010 form MASKLIST, then the current child's s_{umin} is equal to its parent's s_{ymid}. Each child except the first one hit has s_{lmax} equal to the previous child's s_{umin}.

4. Software Implementation

The main data structure used by the HERO algorithm is the octree, which defines bounding volumes for objects contained or partially contained within its terminal nodes. Ray intersections with these objects are calculated separately from the HERO algorithm, so HERO can be applied to any object model for which a bounding octree is constructed (e.g. polygons, half-spaces, quadrics etc.). Each octree node contains a flag to indicate whether or not it is terminal, and if this is true, the node simply contains a pointer to a list of the objects it bounds. Parent voxels contain a list of 8 pointers to child nodes, arranged so that the address of the pointer to child voxel n is equal to the base address of the parent node plus n words. A NIL pointer to a child indicates an empty child node. In addition, each parent node has the positions of its mid-planes (x_{mid}, y_{mid} and z_{mid}) stored to facilitate computation of the distance values s_{xmid}, s_{ymid} and s_{zmid}.

A depth-first search of the octree is required, so that the first intersected terminal voxel returned is known to be the first one hit by the forward-travelling ray. Whenever a parent node's intersected children are exhausted, the algorithm must backtrack one level up the tree and then continue in a depth-first manor. A stack is therefore used for implementation of this search.

One approach to using a stack for the search would be to push information regarding all intersected child nodes (between 1 and 4) for any one parent, entering these in descending order of distance. The top node could then be popped and used for the next level of search, and so on. However, a more efficient approach is to store status information for each parent node on the stack, so whenever a parent is popped, either the next child hit is derived and the parent's status is updated, or if all its children are exhausted, the parent is removed from the stack. In order to implement this operation, the index to MASKLIST should also be stored on the stack, together with a pointer to the previous child's s_{umin} (to be used as the next child's s_{lmax}). Each stack node therefore contains the following:

Pointer to octree node,
s_{lmax}, s_{umin}, s_{xmid}, s_{ymid}, s_{zmid},
CHILDMASK, LASTMASK, MASKLIST [1..3],
MASKLIST_index, Previous_s_{umin}_pointer.

For development purposes, the authors' implementation uses the terminal voxels themselves as simple cubic objects. In order to generate secondary rays (for illumination and reflection) it is necessary to know which face of the cube is struck by the ray. This information takes the form of one further 3-bit mask (SLMASK) carried on the stack, to indicate whether the ray strikes an x, y or z-plane. In conjunction with VMASK, this reveals whether the face struck is $x = x_{min}$, $x = x_{max}$ etc.

The only arithmetic operations required for each parent node (except the root) are computations of s_{xmid}, s_{ymid} and s_{zmid}, plus simple comparisons to generate the various masks. Although it would appear that the new s values could be generated by mid-point interpolation (e.g. $s_{xmid} = (s_{xmin} + s_{xmax})/2$), this is not a reliable technique due to the accumulation of errors. A better solution is to compute the reciprocals of the ray direction components ($1/x_v$, $1/y_v$ and $1/z_v$) once for each ray. These are used in the computation of s_{xmid}, s_{ymid} and s_{zmid} for each parent voxel by the normal ray/plane intersection expression, i.e.

$$s_{xmid} = (x_{mid} - x_E) * (1/x_v) \text{ etc.}$$

Each s value can therefore be calculated by a subtraction and multiplication.

5. Hardware Implementation

Application of parallel processing to ray tracing is highly desirable to minimise image generation time. With almost all ray tracing algorithms, it is easy to conceive a machine which allocates a processor for each new ray, as long as the entire world model is duplicated for each ray processor to avoid memory contention. However, with large world models and bounding octrees, this approach becomes extravagent in its use of memory. Fortunately, the structure of the octree can be exploited to allow other strategies.

At a high level, the HERO algorithm can be considered as a stack-based tree searcher. Distribution of processing without memory contention can be achieved in such a scheme by storing each level of the octree in a different block of memory, each with its own HERO processor. The processor for any one level performs the HERO algorithm, passing its results either up or down to the next tree level processor as appropriate. There is a large amount of message passing in this scheme, but only between adjacent processors. The maximum number of rays which can be processed simultaneously is limited by the depth of the octree, but other strategies for allocation of voxels to processors are also possible.

At a lower level, the close relationship between arithmetic and logical processing in HERO can be exploited in a hardware implementation of the algorithm. A major goal is the minimisation of the amount of floating point processing required, particularly comparisons for generating the various 3-bit masks. If the floating point representation is chosen such that the exponent occupies the most significant bits of the floating point word, and the sign bits of the exponent and mantissa are suitably manipulated, then a floating point number which is larger than another will be represented by a word which has a larger integer magnitude. Simple greater than/less than comparisons of floating point numbers can then be achieved using fixed point comparators.

Sorting of the MASKLIST values into ascending order of referenced s value would appear to be a rather more complex operation, but it can in fact be achieved using only three comparisons of the above type with a few simple logic operations. The boolean results of these comparisons are labelled as follows:-

$$A = (s_{xmid} < s_{ymid}),\ B = (s_{xmid} < s_{zmid}),\ C = (s_{ymid} < s_{zmid}).$$

Using 'X' to represent states which cannot happen, the truth table for the various MASKLIST orderings against A, B and C is as follows:

A B C	MASKLIST[1]	MASKLIST[2]	MASKLIST[3]
0 0 0	1 0 0	0 1 0	0 0 1
0 0 1	0 1 0	1 0 0	0 0 1
0 1 0	X X X	X X X	X X X
0 1 1	0 1 0	0 0 1	1 0 0
1 0 0	1 0 0	0 0 1	0 1 0
1 0 1	X X X	X X X	X X X
1 1 0	0 0 1	1 0 0	0 1 0
1 1 1	0 0 1	0 1 0	1 0 0

By inspection, MASKLIST can be generated by the following logic functions:

Element	Bit 2 (ms)	Bit 1	Bit 0 (ls)
MASKLIST[1]	B NOR C	(NOT A) AND C	A AND B
MASKLIST[2]	B XOR C	NOT(A XOR C)	A XOR B
MASKLIST[3]	B AND C	A AND (NOT C)	A NOR B

Parallel processing can be implemented at a low level within HERO by executing all arithmetic operations concurrently, and all mask generation operations concurrently. Computation of each of the values s_{xmid}, s_{ymid} and s_{zmid} consists of a subtraction and multiplication, thus three parallel arithmetic units are used to generate these values simultaneously. Generation of CHILDMASK, LASTMASK and MASKLIST involves nine arithmetic comparisons, which are achieved by nine integer comparators in parallel. Although current work is aimed at implementation of HERO using floating point processors and logic circuits, there is considerable scope for implementation of these functions on a VLSI integrated circuit (HEROIC).

6. Conclusion

Octree-based ray-tracing is an efficient technique for realistic rendering of complex scenes. The HERO algorithm provides a fast method for determining the nodes of an octree penetrated by a ray in the order they are hit, and is well-suited to VLSI hardware implementation. HERO could be applied to any world model for which a bounding octree can be generated, and hence could form an accelerator for a variety of graphics systems. Applications on which our hardware-enhanced octree ray tracing is targeted include workstations and high quality visual systems for simulation.

7. References

[1] Plunkett, David J. & Michael J. Bailey, "The Vectorisation of a Ray Tracing Algorithm", IEEE Computer Graphics & Applications Vol. 5, no. 8, August 1985, pp 52-60.

[2] Heckbert, Paul S. & Pat Hanrahan, "Beam Tracing Polygonal Objects", Computer Graphics Vol. 18, no. 3 (SIGGRAPH '84 Conference Proceedings), July 1984, pp 119-127.

[3] Whitted, Turner, "An Improved Illumination Model for Shaded Display", Comm. ACM Vol. 23, no. 6, June 1980, pp 343-349.

[4] Kay, Timothy L. & James T. Kajiya, "Ray Tracing Complex Scenes", Computer Graphics Vol. 20, no. 4 (SIGGRAPH '86 Conference Proceedings), August 1986, pp 269-277.

[5] Dippe, Mark & John Swensen, "An Adaptive Subdivision Algorithm and Parallel Architecture for Realistic Image Synthesis", Computer Graphics Vol. 18, no. 4 (SIGGRAPH '84 Conference Proceedings), July 1984, pp 149-158.

[6] Amanatides, John & Andrew Woo, "A Fast Voxel Traversal Algorithm for Ray Tracing", Eurographics '87, August 1987, pp 3-10.

[7] Glassner, Andrew S., "Space Subdivision for Fast Ray Tracing", IEEE Computer Graphics & Applications Vol. 4, no. 10, October 1984, pp 15-22.

[8] Fujimoto, Akira, Takayuki Tanaka & Kansei Iwata, "ARTS: Accelerated Ray Tracing System", IEEE Computer Graphics & Applications, Vol. 6, no. 4, April 1986, pp 16-26.

[9] Peng, Qunsheng, Yining Zhu & Youdong Liang, "A Fast Ray Tracing Algorithm Using Space Indexing Techniques", Eurographics '87, August 1987, pp 11-23.

A VLSI Architecture for Anti-Aliasing

Claudia Romanova and Ulrich Wagner

1. The Aliasing Problem in Computer Graphics

Computer-synthesized images exhibit the typical artifacts of raster displays, called *aliasing, rastering, staircasing* or *the "jaggies"*. Display of an image on a raster CRT requires the sampling the two dimensional image signal $I(x, y)$ to obtain a pixel-based description of intensity. Unfortinately, this sampling process treates the pixel as a mathematical point and the point sampling of an unfiltered object is never correct at any resolution. Aliasing effects (spatial and temporal) are due to undersampling of the image signal. Spatial aliasing occurs when images contain frequencies greater than one half the spatial sampling frequency. Lines that should be straight appear jagged, very small objects may not be visible, portions of long thin objects may disappear.

1.1 Methods for Minimizing Aliasing Effects

There are three basic approaches to remove the aliasing effects:

increase the sampling rate by increasing the display resolution. However, this approach has several limitations, because it also increases the cost of the image production and object edges may fall between pixels at any resolution thus aliasing may occur anyway.

post-filtering: the image is sampled at a higher resolution. Then a digital averaging process is applied to the supersampled image to generate the intensity values in the frame memory. An example is the jagged edges detection and filtering algorithm [6]. The main disadvantage is that details which have been lost during the sampling process cannot been recovered.

pre-filtering or *area sampling* : area sampling treates the image as continuous area of scan lines, where each scan line is a contiguouos band of pixels, each pixel is a square sample, and each sample is of unit edge length. The final pixel intensity is determined by computing the area weighted coverage of all visible objects which cross the pixel. This is equivalent to applying a square area integration filter at a pixel. Visually, this means blurring the picture.

1.2 The Pre–filtering Method

Perfect reconstruction of the original image is still not possible. The pre–filtering algorithms operate by replacing the previous pixel colour value according to the following blending function:

$$C(x,y) \leftarrow \alpha \cdot C_{obj} + (1-\alpha) \cdot C(x,y),$$

where C_{obj} is the colour of the object being drawn, $C(x,y)$ is the previous colour of the pixel (x,y) and α is the response of a low pass filter to the object at (x,y). There are two properties that characterize the pre–filtering algorithm – the filter computation scheme and the filter type. The filter type determines how objects look like. The sampling filter is typically a box, a triangle (tent) or truncated Gaussian. Assigning brightness to each pixel proportional to the fraction of the covered pixel area, is a well established approach. Various implementations have been reported, where the fraction is incrementally calculated [12,21,22,23] or stored in a lookup table indexed by the distance from the object to the pixel [4]. The common property of the above mentioned algorithms is that they are well suitable for vector generation. The other alternative is the approximation of the covered area (α) by using a coverage mask. Each pixel is virtually divided into subpixels, the coverage rate is determined by the number of the subpixels crossed by the object. Section 2 gives a detailed survey of the known anti-aliasing approaches.

1.3 Stochastic Sampling

The last anti-aliasing approach to be mentioned is the stochastic sampling. The image signal is sampled at irregularly placed points and instead of the "jaggies" we get noise. Many feel that this noise is less disturbing for the human observer that the aliasing. The main application area of the stochastic sampling is ray tracing. The investigations in this methods are concentrated on finding good sampling distributions and filtering methods to adaptively increase the sampling rate in regions of the image with high frequencies. A good representation of the stochastic sampling could be found in [9].

1.4 The Role of Gamma Correction

Without accurate gamma correction, anti-aliasing through pixel brightness control is ineffective. Unfortunately, the response of typical video colour monitors is non-linear. Thus, when a linear image is loaded into the frame buffer, a video lookup table is also loaded to correct for the non-linearity of the monitor. The monitor correction function is an exponential function of the form:

$$\text{lookup value} = \text{intensity}^{1.0/\gamma}$$

Gamma(γ) represents the non-linearity of the monitor and usually is in the range 2.0 to 3.0. Another problem is that an increment of 1 in the image intensity is mapped into a much larger increment for the display. Thus, the available resolution of the display system in the lower intensity range is not being used [2]. One solution of the problem is to use a lookup table with logarithmic mapping of 12-bit input to 8-bit output values.

The recalculating of the display-data values by compensation tables [5] is one choice for gamma correction.

2. Review of Existing Anti–Aliasing Techniques

All algorithms mentioned in this section, except the cited subpixel coverage mask methods, were implemented by the authors for a straight line drawn from a pixel centre at a gradient 5/7. It is to point out, that only the using of a coverage mask allows several objects per pixel (more than two). All algortihms can be modified to antialias polygon edges, too.
The task of the anti-aliasing process is to *determine the covered area as exact as possible.* Two main approaches can be distinguished.

2.1 Exact Computing of the Covered Area

The most exact algorithm for calculating the intersected area is the method by *Pixel Integration*, described in [22]. The anti-aliased lines drawn with the Field's algorithms [12] have a width less than one pixel and look on the raster display thinner, but the area crossed by the vectors is determined exactly, too. These methods work incrementally beginning with the vector start point, i.e. they are *object oriented* and take into consideration three pixels per column. Algorithms for drawing anti-aliased objects on raster output devices are presented in great detail in [11].

2.2 Approximate Computing of the Covered Area

All methods determining the intensity independently of the pixel location can be indicated as *pixel oriented.*

2.2.1 Using the Distance

The following methods [16,21] use the fractional part of a coordinate value of the boundary edge position as a criterion for pixel intensity determination. The algorithm of Fujimoto and Iwata [14] calculates the distance between the vector and the pixel. In both approaches the pixel intensity is inversely proportional to the calculated distance. In [23] the control parameter of the Bresenham's algorithm is interpreted as a measure of the distance from the straight line to the pixel centre. This algorithm is known as modified Bresenham's algorithm. These algorithms involve two pixels per column.
The algorithm cited by Gupta and Sproull in [17] uses the perpendicular distance from the pixel centre to the edge as index into a lookup table with precomputed conic filtering function to determine intensities at the pixels, but it requires seperate treatment for the end points of the vector. The above algorithms rely upon incrementally computations, too. Since in our system architecture the same distance is available, this approach could be used in the Filtering Stage, but it has several disadvantages, still to be discussed.

2.2.2 Using the Coverage Mask

The methods presented in [3,13,19] and in this paper compute an anti-aliased image using subpixel mask. The bitmask can be interpreted as a 2D array of subpixels, each

subpixel being off or on depending on the intersected area of the object with the pixel. The quantized area is computed by summing the number of bits that are on in the bitmask and dividing by the total number of bits in the mask. So, this fractional coverage multiplied by the object intensity determines the final intensity of the displayed pixel. The coverage mask can be either generated or determined by indexing a precomputed table of bitmasks. Our algorithm is similar to the well known "A–buffer" algorithm [20]. The authors of [1] proposed the "gz-buffer". The gz-buffer can be split into an usual z-buffer and a g-buffer containing the geometrical information g for the pixel. Each pixel is divided into sixteen subpixels, therefore the z-buffer needs only four bits per pixel. The approach in the mentioned paper uses the oversampling process for anti-aliasing and in the second one [15] the modified Bresenham's algorithm is applied.
On the Workstation PS 390 of Evans & Sutherland a real-time anti-aliasing technique in hardware is already implemented, which uses 8 x 8 subpixel resolution and produces anti-aliased vectors of high quality.

2.2.3 Compositing Technique

According to this technique the image is described by a quadruple (r, g, b, α), where α is called alpha channel and indicates the covered fraction by the object. The α value lies in the interval [0.0,1.0], where 0.0 corresponds to no coverage and 1.0 to full coverage, the fractions mean partial coverage. The Duff's approach in [10] combines the α compositing algebra with z-buffer to provide simple anti-aliasing.
Assuming a quadratic pixel model the z-values of the object (foreground) at the four pixel corners are computed and compared with these of the previous object (background). Then the total area of both objects is found and the pixel colour values are altered by this value. The (r, g, b, α, z) representation for more than two objects assumes that they are either adjacent in depth or requires knowledge of their front-to-back ordering. The method fails by two elements sharing an edge.

2.3 Error Estimations of Quantized Area Sampling Technique

The difference between the exact and approximated area depends both on the line slope and the chosen subpixel resolution (N). The quantized pixel area (α) can be expressed as:

$$\alpha = \frac{\sum_{i=1}^{N} SubpixelValue_i}{N^2} \qquad SubpixelValue_i \in [0,1]$$

The worst case corresponds to horizontal and vertical lines, which cross the subpixel centers and the maximum error is equal to $\frac{1}{2N}.100\%$. Statements about errors of other anti-aliasing algorithms are scarce, but when they are given, the errors lie in range of $8 - 12\%$.

3. The PROOF System for Computer Image Generation

The idea of the architecture of our system PROOF (Pipeline for Rendering in an Object Oriented Framework) is based on an object space subdivision approach, given by Cohen [8] and extended by Weinberg [27] and Straßer [26]. Figure 1 shows a block diagramm of the overall system. The geometric processor executes the perspective and geometric transformations, the clipping and the triangulation of the objects in the scene. Each processor is dedicated to an object primitive in the image space. The object processors are organised in pipeline (Object Processor Pipeline). The Look-Up-Tables between the stages are loaded with parameters neccessary for the following Shading and Filtering Stages. The system operates in two modes: *load* and *pixel.*

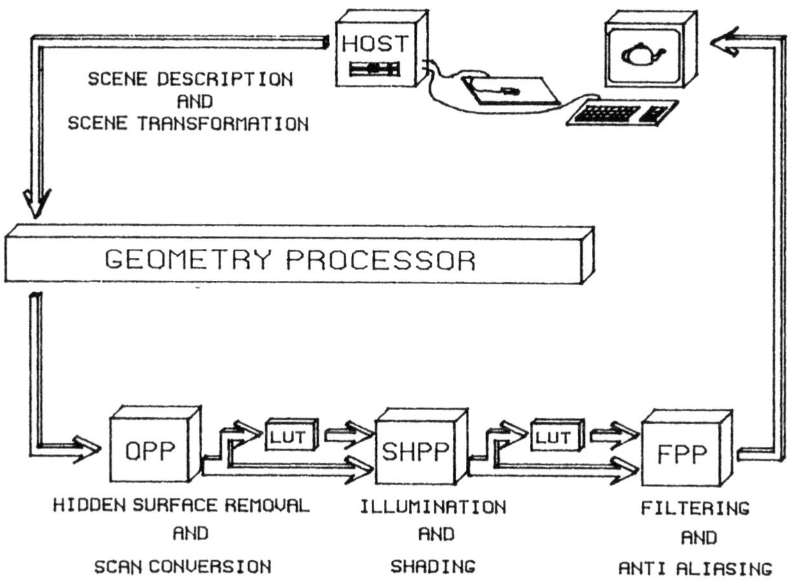

Figure 1: The Architecture of PROOF

3.1 Load Mode

The same pipeline is used for loading the object data. In the load mode the object processors, the shading processors and the LUT's are preloaded by the geometric processor with specific properties describing the objects and the scene. For example, the first LUT contains parameters needed by the Shading Processor Pipeline such as parameters referring to the position and colour of the light sources in the scene and surface properties of the objects. The second one is loaded with geometrical informations required by the Filtering Stage. Identifying each object from the scene by an object number (object identifier) and using this as address into the LUT's minimizes the bandwidth of the dataflow in the system essentially.

3.2 Scan Conversion and Hidden Surface Removal Algorithms

Pixel computations consist of the scan conversion of the object and the interpolation of the Z coordinates and colour values. The scene to be rendered consists of triangles as objects primitives. A triangle edge is given by its bounding line, described by the

Hessian normal form as follows:

$$g \ : \ \cos\theta.x + \sin\theta.y - p = 0$$

The values of $\cos\theta$, $\sin\theta$ and p are determined from the end-points of the line, traversed in an anticlockwise direction, so that the triangle will always lie in the left halfplane of the line.
The scan conversion algorithm applied in the OPP can be classified as *inside testing* algorithm and takes advantage of the method of differences. Every object processor calculates the distance between each edge of its own object and the centre of a pixel (x,y) by the equation:

$$d_i(x,y) \quad = \quad \cos\theta_i.x + \sin\theta_i.y - p_i \qquad (1)$$

for $i = (1,2,3)$. Evaluating the value of d_i for every pixel requires two multiplications and two additions. Fortunately, the calculations can be made incrementally with only one addition per step. The value of $d_i(x,y)$ for successive pixels is obtained by initializing a register for each edge with the value of $(-p_i)$ and adding $\cos\theta_i$ for each step along the x axis and adding $\sin\theta_i$ for each step along the y axis. The basic rendering operations are as follows:

command	execution
new frame	$d_i(0,0) = -p_i$
new pixel	$d_i(x+1,y) = d_i(x,y) + \cos\theta_i$
new scan line	$d_i(0,y+1) = d_i(0,y) + \sin\theta_i$

The value of d_i will be negative for all pixels that are on the left side of the triangle edge and positive for all pixels that are on the right side of the triangle edge. If any d_i is in the range $[-R,+R]$ (R denotes the pixel radius and for a quadratic pixel with unit length is equal to $\frac{\sqrt{2}}{2}$), the pixel is partially covered from the object. If all d_i are greater than $+R$, the object covers the pixel totally and if at least one d_i is less than $-R$ and another d_i is (are) greater than $+R$, then the pixel is completely outside the polygon.
In addition to the object coordinates at each vertex of the object, its depth and its colour are stored in the object procesor. It generates point samples of these values for each pixel within the triangle by linear interpolation between the vertex values incrementally in a manner similar to the calculation of the function for the distance (1). The hidden surface removal approach in the OPP relies upon the depth-buffer algorithm, more exact the distributed z-buffer. For each pixel a list with objects is maintained, which cross the pixel. The next sections explain briefly the architectures of each stage of PROOF.

3.2 Object Processor Pipeline

All object processors have the same architecture design, shown in Figure 2, and each object processor consists of primitive processor, comparator, FIFO and control logic. The primitive processor scan converts its object and performs the interpolation of the depth and normal colours vectors. The seven update units work in parallel to achieve high performance. The comparator is responsible for the decision if the interpolated z value of the local object lies further back than the objects' depth values stored in the list. The FIFO receives the data sent from the last object processor and serves as a buffer between the singles object processors. During the scan conversion, each object

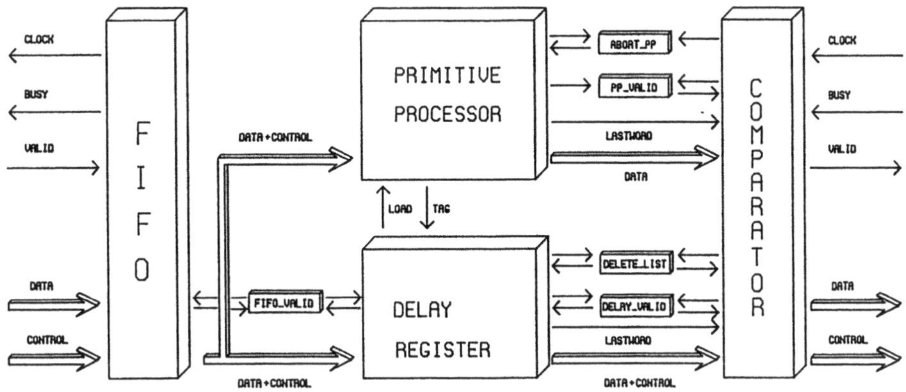

Figure 2: The Internal Design of the Object Processor

processor receives from the previous processor a depth sorted list of potentially visible objects for the current pixel. If the object processor's own object lies within the triangle and the z value is less than that for the triangle computed by preceding object processor for this pixel, the local object is inserted in the list. Else if the object totally hides the previous candidate objects and is non-transparent, the hidden objects are removed from the list. At the end of the object processor pipeline a list of objects for the current pixel emerges.

3.3 Shading Processor Pipeline

Afterwards, these data pass a multiple light source shading processor pipeline (SHPP) in which a Phong–like shading is performed. The illumination model implemented in the SHPP includes terms for ambient light, diffuse and specular relection. The calculations are implemented such, that only a single type processor is needed. The Shading Processors are connected in a series to form a pipeline, too. The pipeline is initialized with the ambient term. For each light source in the scene two shading processors are calculating the other two terms. A review about the investigations in reducing the Phong Shading method is given in [7].

3.4 Filtering and Anti-aliasing Stage

The last stage is the filter processor pipeline (FPP) which for each potentially visible object from the list re-scans it over a high resolution subpixel mask and applies a predetermined filter function over the visible portions of the object. Finally, these sums are multiplied by the object's colour. The end colour sums are sent to the frame buffer.

Subpixel position information is required by Filtering Stage to properly antialias an object. The key elements of our system are described in detail in [24] and [25].

4. Algorithm for Filtering

In order to generate the subpixel coverage mask of an object we need geometric information about object intersection points with the pixel area. The object oriented anti-aliasing algorithms are to be rejected, because we would need a filter processor per object, too. That would be too expensive. The disadvantage of the method based on the distance is that it is not applicable for more than two objects per pixel. The only apparent way to solve this problem is subpixel coverage mask generation.
The algorithm in the Filtering Stage is performed on a per-pixel basis so as to preserve the pixel area coherence. It makes use of the distance calculated in the OPP. Our task is to determine for every object of the list how many subpixels from the object are crossed. We calculate then the perpendicular distances between each subpixel centre and the triangle edges. Since the function $d_i(x,y)$ is linear, we can obtain the values

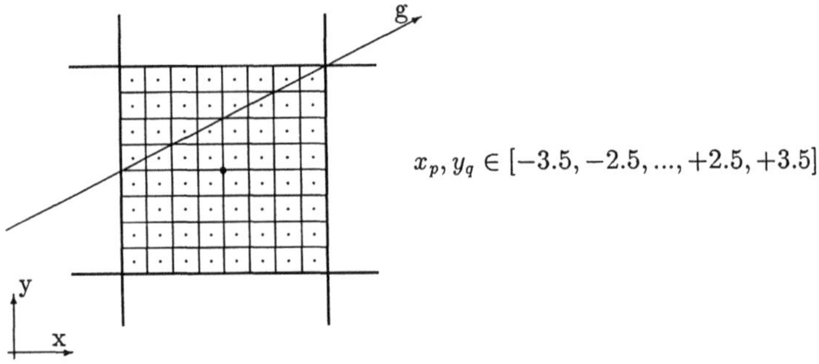

Figure 3: Pixel Area with 8 x 8 Subpixel Resolution

only with adding the first forward differences of the function for n-step change in x- and y-direction. The distance between the triangle edge and the subpixel (p,q) within the pixel area can be expressed as:

$$d_{pq}(x,y) \;=\; \cos\theta_i.(x.sp + x_p) + \sin\theta_i.(y.sp + y_q) - p_i, \tag{2}$$

with the following meanings:

x, y	:	x– and y–coordinates of the pixel
sp	:	subpixel resolution
x_p, y_q	:	offsets in x– , y–direction within the pixel area
$\cos\theta_i$, $\sin\theta_i$, p_i	:	edge parameters in Hessian normal form
$d_{pq}(x,y)$	:	distance from subpixel centre (p,q) to the edge

The equation (2) can be simplified to:

$$d_{pq}(x,y) \;=\; \underbrace{\cos\theta_i.x.sp + \sin\theta_i.y.sp - p_i}_{d_i(x,y)} + \cos\theta_i.x_p + \sin\theta_i.y_q \tag{3}$$

$$\;=\; d_i(x,y) + \cos\theta_i.x_p + \sin\theta_i.y_q \tag{4}$$

Thus, the desired distance can be obtained only by adding or substracting appropriate coefficients to the value of $d_i(x,y)$, namely multiples of $\cos\theta_i$ and $\sin\theta_i$. Figure 3 shows the pixel area and the corresponding coefficients. All bits in the coverage mask for these the calculated subpixel distance $d_{pq}(x,y) \leq 0$ are to be set.
The object list is depth sorted with respect to the pixel centre. The assumption, that this list is correctly sorted for each subpixels too, would cause visible errors, as the following example illustrates. If there are two objects in a pixel, one covering the pixel totally, another one, lying behind, but having greater z gradient and not covering the center of the pixel. Since the z-values in the OPP are always computed at the pixel center, it can happen that the interpolated z-value of the second object is smaller than the z-value of the first one. This causes the second object to be visible, in spite of being hidden by the totally covering object. For this reason we need a z-buffer within a pixel region, too. Analogous to the determination of the subpixel distance we can calculate the needed depth values at subpixels as follows:

$$z_{pq}(x,y) \quad = \quad z(x,y) + A_z.x_p + B_z.y_q \tag{5}$$

For transparent objects we suppose, that they are correctly sorted in the list.
Since the PROOF-architecture supports translucency as object property, a transparency modul in the Filtering Stage is integrated too. Generally, assuming that n transparent objects each with colour C_{obj_i} and transparent factor t_i cross a given pixel, the final pixel colour is to be determined by the equation $C_{pixel} = \sum_{i=1}^{n} C_i + C_{bg}$. The bg denotes the background in the scene with $t_{bg} = 0$ and each term C_i is expressed as:

object colour	remaining term
$C_1 = (1-t_1)\cdot C_{obj\_1}$	$R_1 = 1 - t_1$
$C_2 = (1-R_1)\cdot(1-t_2)\cdot C_{obj\_2}$	$R_2 = R_1 + (1-R_1)\cdot(1-t_2)$
$\vdots$	$\vdots$
$C_n = (1-R_{n-1})\cdot(1-t_n)\cdot C_{obj\_n}$	$R_n = R_{n-1} + (1-R_{n-1})\cdot(1-t_n)$
$C_{bg} = (1-R_n)\cdot C_{bg}$	

Fortunately, only one multiplication depending on the current transparent factor (t_n) and the remaining term (R_{n-1}) is needed. The same approach can be used at subpixel level, as described later. By test and display of different images we found, that 8×8 subpixel resolution in the xy plane and 4×4 subpixel resolution for the depth value seem to be reasonable. The algorithm antialiases images of wire-frames as well as triangles. The algorithm handles adjacent and intersecting polygons thanks to the geometric information given by the object parameters $\cos\theta_i$ and $\sin\theta_i$.

5. Architecture Design of the Filtering Stage

The Filtering and Anti–Aliasing Stage (FAAST) is divided into three distinct processing stages: d–, z–update units (hereafter Update Units) and subpixel processor area (SPA) (Figure 4). The internal hardware structures of the Update Units and SPA are quite different, being designed for their dedicated tasks. Since neither feed-back of data nor that of control from the SPA to the Update Units is required, the operations of the SPA might be completely overlapped in a pipelining manner with that of the Update Units, which are then processing the next object.

5.1 The d– and z–Update Units

Calculating the multiples of $\cos\theta_i$ and $\sin\theta_i$ (A_z and B_z for the depth value) can be done easily with a binary tree of adder. Figure 5 illustrates a binary-tree multiplier, as used in the Pixel Planes System. Each node of the binary tree has two inputs (top and side) and two outputs (left and right), which are bit-parallel and synchronous with the system clock. The value of the left output is the value of the top one delayed by one clock cycle, the right output is the sum of the top and side inputs. The value at the side input depends on the node level in the tree and is multiple of a and a power of 2. The last tree stage delivers the inverted value of the input, too. For a pixel grid of size $N \times N$ subpixels (assuming N to be power of 2), the tree has $n = \log_2(N/2)$ levels and the value of each side input is $2^n.a$, where n-th level corresponds to the top node of the tree. The total number of the required additions in the tree is $\sum_{m=1}^{n} 2^{n-m}$.

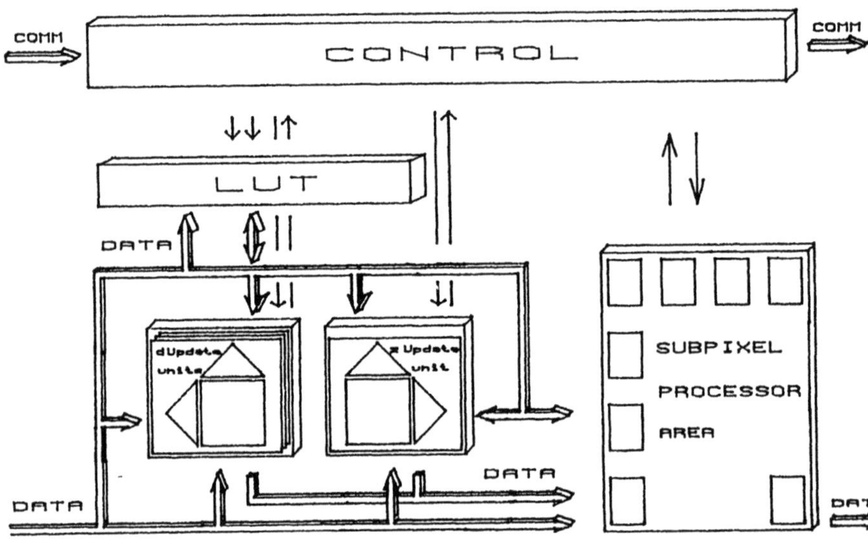

Figure 4: The Architecture of FAAST

The Update Units are similar in their architecture and the task of them is the generation of the subpixel coverage mask. The d–Update Units take advantage of the distance calculated in the OPP, as described earlier, and using the edge parameters loaded in the LUT calculate for each subpixel of the pixel area the distance from the subpixel centre to the object edge. So the 2-dimensional subpixel coverage mask in xy plane can be generated for each object in the sorted list. The task of the z–update unit is similar, but we have to determine the z-values within the pixel area, here. Each type Update Unit, whose block diagramm is shown in Figure 6, is built from two trees and an array of identical Processing Elements (PEs). For a $N \times N$ subpixel resolution we need N^2 PEs for each Update Unit. A PE itself is composed of (de)multiplexing devices, registers

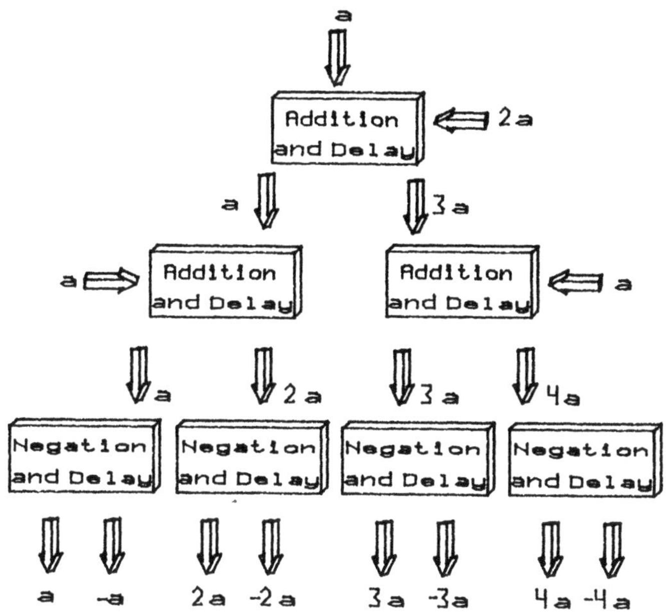

Figure 5: The Binary-Tree Multiplier

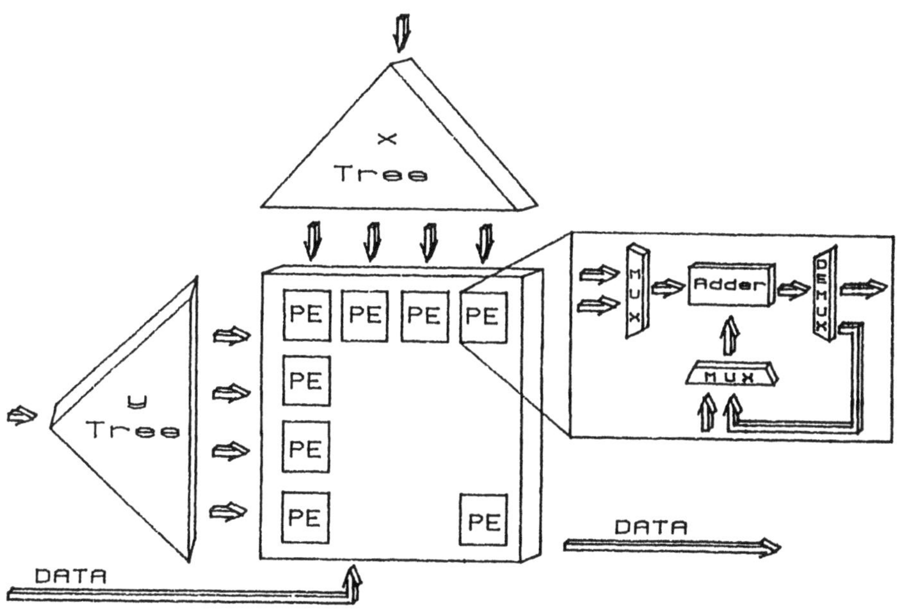

Figure 6: The Update Unit

and an adder. The structures of the trees in the Update Units differ in the number of needed adders and in the bit-width of the addition. The multiplier tree takes in an N-bit data and outputs $(N+3)$-bit-long result. The following table shows the difference between the trees in both types of Update Units, assuming that all adders in the tree have the same bit width:

tree in	number of		bits of	
	levels	adders	input	adder
d-update unit	2	3	16	19
z-update unit	1	1	36	39

The output of the d-Update Units is the 64 bit subpixel coverage mask resulting from the logical AND of the coverage masks for each edge of the triangle. And the results of the z–Update Unit are the depth values for 4×4 subpixel resolution within the pixel area.

5.2 The Subpixel Processor Area

The SPA operates as a single instruction, multiple data (SIMD) processor, executing each instructions on all subpixel processors at the same time. Sixteen Subpixel Processors (Figure 7) constitute the Subpixel Processor Area and each of them consists of registers for object colour, for the transparent factor and for the object depth value at the subpixel and four Subpixel Processor Units (Figure 8). The outputs of the SPU are the colour components for all objects at the current subpixel (sp_rgb_out), which are processed through two stages of adders. Additional summations are made outside of the SPP in a sequence of four adders stages. For each covered subpixel of the object mask the input z-value of the object at that point is compared to the value stored in the z-register. If the comparison is successful, the three colour component values of the object are copied into the colour registers and the z-register is updated with the input z-value. At the end of the procedure all colour components for the actually pixel are summed, divided by the total number of the subpixels and sent to the frame buffer. The algorithm in the Filtering and Anti-aliasing Stage can be described by the following C-like functions:

```
/*              anti-aliasing algorithm in FAAST                         */
/*                                                                       */
  for each pixel(x,y) do
    { null sp_rgb_out, final_rgb;
      initilize sp_z;
      for every object from object_list do
       { null obj_mask, sp_rgb;
         determine obj_mask and obj_z;
         spu(); }
      add all sp_rgb_out to final_rgb; }
/*                                                                       */
/*    spu realizes the tasks of the subpixel processor unit              */
/*                                                                       */
spu()
{
 if (obj transp) and (all prev obj transp) and (subpixel set) do
```

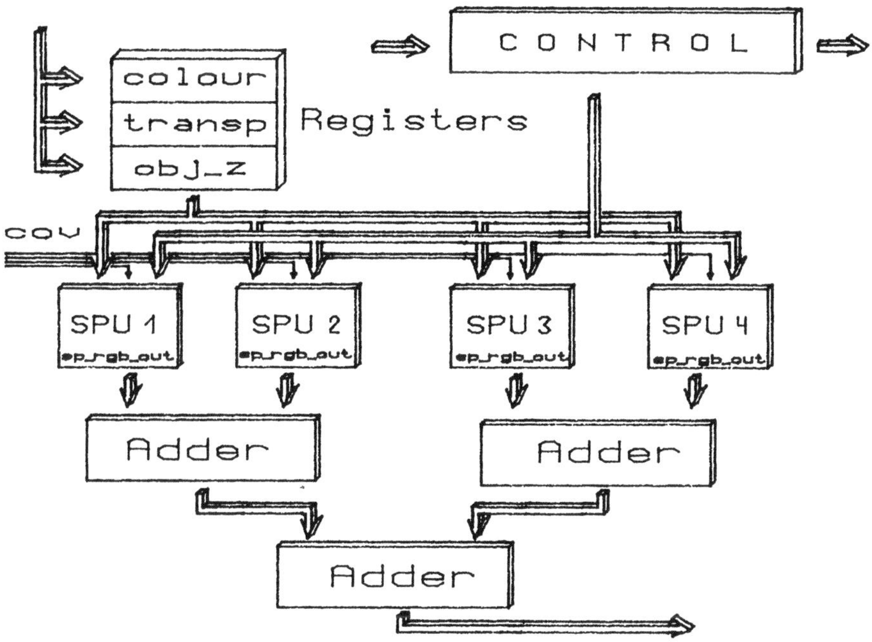

Figure 7: The Structure of the Subpixel Processor

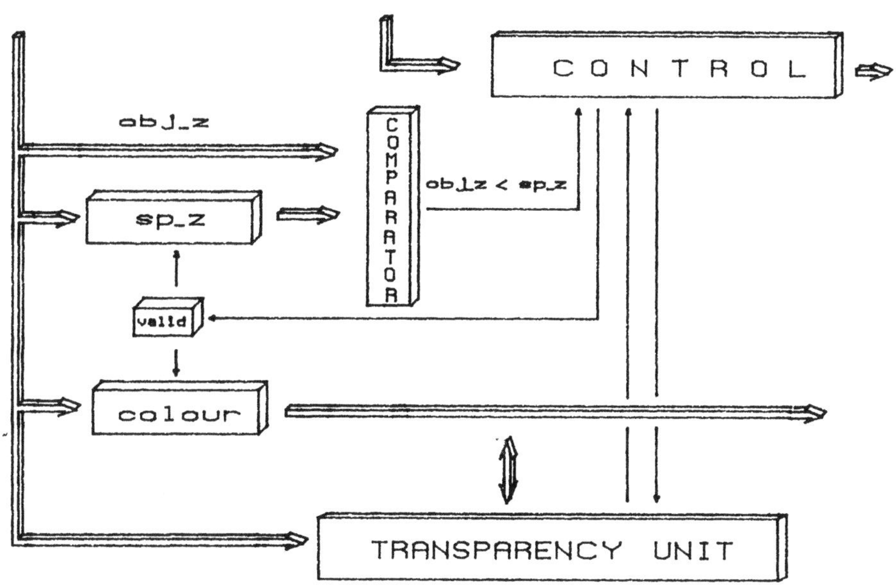

Figure 8: The Design of the Subpixel Processor Unit

```
   { activate transp_unit; }
  if (obj non transp) and (subpixel set) and (obj_z < sp_z) do
   { move obj_rgb to sp_rgb;
      move obj_z   to sp_z; }
  if (last obj) and (subpixel not set) do
   { move obj_rgb to sp_rgb;
     determine sp_rgb_out; }
}
```

5.3 The Lookup Table and Control Unit

The parallel processing manner of the Update Units allows, that the LUT's can be divided into two parts and each part is connected to the Unit by data bus with defined bit width. All Control Units can be realized as PLAs.

6. System Performance Estimation

The pixel processing time is limited by the longest updating time of any stage in the whole pipeline. We restrict our considerations to the processing time in the FAAST. The Update Units process the objects from the list successively and the SPA needes data for all objects in order to calculate the final pixel colour components. The processing time of the Update Units is more decisive than that of the the SPA. Supposing that there are three objects in the list for the current pixel, the Update Units have to determine for two objects (the third one is the background and covers the pixel completely) the subpixel coverage mask, i.e. $t_{pixel} \leq 2.t_{UpdateUnit}$. The processing in the Update Unit includes:

activity	operation	number of needed clock cycles
fetching the object parameters	access to the LUT	1
calculating the coefficients	2 additions and negation	3
determine the d_{pq}	2 additions and evaluate the sign of d_{pq}	3

The main operation to be performed in the Update Unit is addition. The carry–skip adder presented in [18] is well suited for addition of data with large width, e.g. the addition of two 20-bit binary numbers is performed in $10T$ (T is the gate delay). The negation operation needes no extra time, because it can be included in the following addition by setting the input carry to one. So, assuming that the T is $2ns$ the total processing time of an Update Unit is approximately $150ns$ and the $t_{pixel} = 300ns$.

7. Conclusions

Anti-aliasing is needed to preserve high quality images. The described algorithm and architecture to solve this problem has been dictated by the PROOF architecture. The proposed solution seems to be realisable with today's technology. It demonstrates, that anti-aliasing in real time system is feasible.

References

[1] Beigbeder, M., Ghazanfarpour, D., Peroche, B.: The "GZ-Buffer" Method for Antialiasing. International Electronic Image Week, April 1986.

[2] Blinn, J.F.: Dirty Pixels. IEEE Computer Graphics and Applications, 100–105, July 1989. Jim Blinn's Corner.

[3] Carpenter, L.: The A-buffer, an Antialiased Hidden Surface Method. Computer Graphics, 18(3):103–108, July 1984.

[4] Catmull, E.: An Analytic Visible Surface Algorithm for Independent Pixel Processing. Computer Graphics, 18(3):109–115, July 1984.

[5] Catmull, E.: A Tutorial on Compensation Tables. Computer Graphics, 1–7, 1979.

[6] Chryssafis, A.: Anti-Aliasing of Computer - Generated Images: A Picture Independent Approach. Computer Graphics Forum, 5:125–129, June 1986.

[7] Claussen, U.: On Reducing the Phong Shading Method. In F.R.A. Hopgood and W. Straßer, editors, EUROGRAPHICS'89, pages 333–344, Eurographics Association, Elsevier Science Publishers B.V. (North-Holland), 1989.

[8] Cohen, D.: A VLSI Approach to the CIG Problem. 1980. Presentation at SIGGRAPH 1980.

[9] Dippe, M.A.Z.: Anti-Aliasing through Stochastic Sampling. Computer Graphics, 19(3):69–78, July 1985.

[10] Duff, T.: Compositing 3-D Rendered Images. Computer Graphics, 19(3):41–44, July 1985.

[11] Field, D.: Algorithms for Drawing Simple Geometric Objects on Raster Devices. PhD thesis, Princeton University, Juni 1983.

[12] Field, D.: Two Algorithms for Drawing Anti-Aliased Lines. Graphics Interface, 87–95, 1984.

[13] Fuime, E., Fournier, Al.: A Parallel Scan Conversion Algorithm with Anti-Aliasing for a General-Purpose Ultracomputer. Computer Graphics, 17(3):141–150, July 1983.

[14] Fujimoto, A., Iwata, K.: Jag Free Images on a Raster CRT. In Computer Graphics, Theory and Applications, pages 2–15, Springer-Verlag, Tokyo Berlin Heidelberg New York, 1983.

[15] Ghazanfarpour, D., Peroche, B.: A Fast Anti-Aliasing Method with a Z-Buffer. In G. Marechal, editor, EUROGRAPHICS'87, pages 503–512, Eurographics Association, Elsevier Science Publishers B.V. (North-Holland), 1987.

[16] Guangnan, N., Tanner, P., Wein, P., Bechthold, Gr.: An Algorithm for Generating Anti-Aliased Polygons for 3-D Applications. Graphics Interface'83, 23–32, 1983.

[17] Gupta, S., Sproull, R.F.: Filtering Edges for Gray-Scale Displays. Computer Graphics, 15(3):1–5, August 1981.

[18] Guyot, A., Hochet, B., Muller, J.M.: A Way to Build Efficient Carry-Skip Adders. IEEE Transactions on Computers, C-36(10):1144–1152, October 1987.

[19] Hoffert, E.M., Bishop, G.: Exact and Efficient Area Sampling Techniques for Spatial Antialiasing. December 1985. Technical Memorandum, AT & T Bell Laboratories.

[20] Kedar, At.: Enhancement and Implementation of the A-buffer Rendering Algorithm. Master's thesis, University of Regina, Department of Computer Science, Regina, Saskatchewan S4S OA4, Canada, February 1987.

[21] Ketcham, R.L.: A High-Speed Algorithm For Generating Anti-Aliased Lines. Proceedings of the SID, 26(4):329–336, 1985.

[22] Pitteway, M.L.V., Olive, P.M.: Filtering Edges by Pixel Integration. Computer Graphics Forum, 4:111–116, 1985.

[23] Pitteway, M.L.V., Watkinson, D.J.: Bresenham's Algorithm with Grey Scale. Communications of the ACM, 23(11):625–626, November 1980.

[24] Schneider, B.-O.: A Processor for an Object-Oriented Rendering System. Computer Graphics Forum, 7:301–310, 1988.

[25] Schneider, B.-O., Claussen, U.: PROOF: An Architecture for Rendering in Object Space. In A.A.M. Kujik, editor, Advances in Graphics Hardware III, Eurographics, Spinger, Berlin, to appear in 1989

[26] Straßer, W.: A VLSI-oriented Architecture for Parallel Processing Image Generation. In G.L. Rejins and M.H. Barton, editors, Highly Parallel Computers, pages 247–258, Elsevier Science Publishers B.V. (North-Holland), 1987.

[27] Weinberg, R.: Parallel Processing Image Synthesis and Anti-Aliasing. Computer Graphics, 15(3):55–61, August 1981.

PS: Polygon Streams
A Distributed Architecture for Incremental Computation Applied to Graphics

Rajiv Gupta

Department of Computer Science
California Institute of Technology
Pasadena, California 91125
USA

Abstract

Polygon Streams is a distributed system with multiple processors and strictly local communication. A unique custom VLSI chip that constitutes an independent processing module forms a stage of the PS pipeline. The number of these modules in PS is a variable that is determined by the application. PS features a modular architecture, multi-ported on-chip memory, bit-serial arithmetic, and a pipeline whose computation can be dynamically configured. The PS design closely subscribes to the system characteristics favored by VLSI.

The task of scan conversion for rendering computer graphics images on raster scan displays is very intensive in computation and pixel information access. It is very coherent and suitable, however, for forward difference algorithms. The discrete and regular layout of the raster display, in conjunction with the largely local effect of a pixel on an image, make rendering amenable to parallel architectures with localized memory and communication. These are precisely the attributes favored by VLSI and typical of PS.

A modification of the Digital Differential Analyzer (DDA) is implemented to Gouraud Shade and depth buffer convex polygons at high speeds. The scan conversion task is distributed over the processors to efficiently subdivide the image space and maximize concurrency of processor operation.

A study of the tradeoffs and architectural choices of the PS reveal the merits and deficits of the PS approach in comparison with Pixel-Planes, SLAMs, Super-Buffers, and SAGE.

1 Introduction

Computer Graphics has traditionally subscribed to two primary goals: realism and speed. The first goal attempts to conceive, create, and display scenes that are true to life. The latter goal is devoted to practicality. We wish to retain as much of the realism as possible as we make the modeling and rendering of images faster. Interactive, realistic computer graphics is our ultimate goal.

In the near future, as the scenes that we want to create get more complex and the phenomena that we want to simulate get more involved, we will continue to lack the ability to manipulate realistic images in real-time. To justify spending the time and computational resources in obtaining an exact image, we should first convince ourselves that the interactions in the scene – of objects, viewing parameters, light sources, to name a few – are as we had envisaged them. In order to make this decision with reasonable accuracy we need a way to quickly preview approximations of these scenes interactively. Real-time simulations and animation also require images containing thousands of polygons to be rendered within a frame time. The work described in this thesis renders shaded and depth-cued polygons on raster scan displays at speeds that match the requirements of these applications.

1.1 Organization of the Thesis

In Section 2 we present the context of the PS System application in a graphics pipeline. In Section 3 we discuss the system architecture of PS. This includes the VLSI implementation of the incremental algorithm discussed above and a detailed design of the scan line interpolation subsystem (SLI) of PS. In Section 4 we discuss some of the salient features of the PS design. Section 5 compares the PS architecture with that of the Pixel-Planes, SLAM, Super-Buffers, and SAGE systems. In Section 6 we discuss enhancements to PS and explore the realm of application possibilities of the architecture. Section 7 concludes the document. Appendix A is a report on the fabrication details and expected performance of the PS system.

2 The Graphics Application: The Depth-Buffer Rendering Pipeline

In the graphics pipeline for converting an image from the world or modeling coordinates to the screen or pixel coordinates, a polygonal object is typically described in terms of the position, x, y, and, z, and color, r, g, and b, attributes of its vertices. The viewing, clipping, perspective, and lighting computations give us a new set of vertex and color attributes in screen space. The PS system accepts the screen coordinate and color information of the vertices of a convex polygon and interpolates them to produce (z, r, g, b)-tuples for all pixels that lie within the polygon. This information is suitable for raster-scanned displays. The shading and visible surface determination task is the most computation and communication intensive component of the graphics pipeline. We render convex polygons at very high speeds by implementing two of Computer Graphics' well-known algorithms: depth-buffering [Foley 84] and Gouraud shading [Gouraud 71].

3 The Architecture and Organization of the PS System

3.1 System Architecture

The system performs two distinct but dependent interpolations: along polygon edges and along scan lines. Along polygon edges we interpolate x, r, g, b, and z against y to obtain pixel values at the edge pixels that constrain the polygon on consecutive scan lines. Along a scan line we interpolate r, g, b, and z against x to obtain pixel values at consecutive pixels that lie within the edge pixels of that scan line.

For a given scan line, only after the values at the edge pixels have been calculated may the values for pixels intermediate to those edge pixels be calculated. Note that scan line interpolation is dependent on the polygon edge interpolation only for the values of the edge pixels. Also, interpolations along different scan lines are mutually independent for the same reason. This lends the system to a functional hierarchy – polygon edge interpolation and scan line interpolation. The PS system architecture is shown in Figure 1.

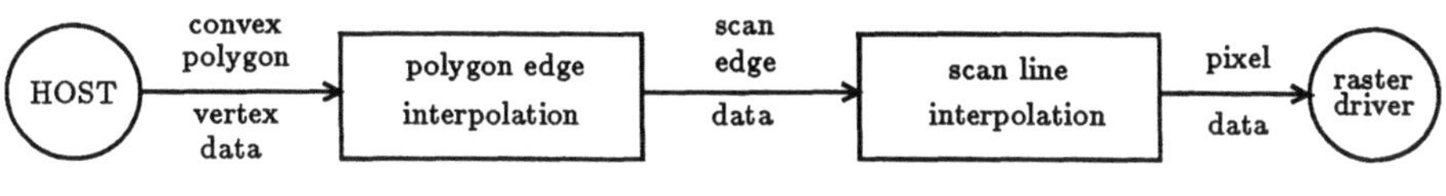

Figure 1: The PS System Architecture

Since we are not constrained by any polygon processing order we can arbitrarily extend the architecture at this level. If the polygon edge interpolation (PEI) is the bottleneck we can balance the system with multiple PEIs. Conversely, if the scan line interpolation (SLI) is the bottleneck we can balance the system with multiple SLIs. A hybrid approach, allows us to optimally balance the number of SLI's with respect to the PEI's, and the numbers of PEI's and SLI's with respect to cost-speed-accuracy indices. It is now trivial to fine tune the architecture to the scene composition. For example, scenes with consistently small polygons, as in the case wherein the accuracy in rendering complex curved surfaces is enhanced by fracturing the surface into larger numbers of smaller polygonal artifacts, require more PEIs.

The potential for parallel hardware is readily apparent. This paper deals with the detailed design and implementation of the scan line interpolator subsystem.

3.2 The Scan Line Interpolation Subsystem

3.2.1 Parallelism

Since the interpolations along different scan lines are independent operations, the scan line interpolator subsystem consists of several scan line interpolators which work in parallel. The best that we can do is to have an independent interpolator for each scan line. The time to interpolate the entire convex polygon is then the time to interpolate its longest scan line extent.

3.2.2 Pipelining

Interpolations along the same scan line for different polygons are mutually independent. Also, the interpolation for a scan line can be broken into discrete and repetitive steps that successively interpolate for the pixels that lie on the scan line. Thus each scan line interpolator can be pipelined so that interpolation for polygon '*a*' can be started before that for polygon '*b*' is completed.

For a given scan line y, we may interpolate at $(x1, y)$ for polygon '*a*' concurrently with the interpolation at $(x2, y)$, $x1 \neq x2$, for polygon '*b*'. Note that $(x1, y)$ and $(x2, y)$ cannot belong to the same convex polygon because the interpolation at $max(x1, x2)$ is dependent on the cumulative error calculated at the interpolation of $min(x1, x2)$ if $x1, x2$ belong to the same polygon. Thus the scan line processor can be split into many sub-scan line processors each responsible for a set of m unique pixels and potentially interpolating for a different convex polygon. In Figure 2 both convex polygons are interpolated in parallel by a non-intersecting set of sub-scan line processors.

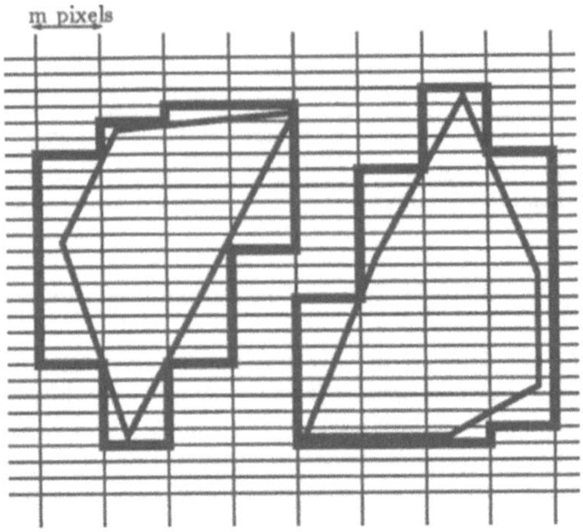

Figure 2: The Parallelism and Pipelining in the system: both polygons are interpolated in parallel by the pipelined scan line processors.

The sub-scan line processor stores its pixel-set information, $(z, r, g, b)_i$ for $1 \leq i \leq m$, in its private memory. This memory is organized as a pixel and depth buffer for the sub-image. Thus over the array of processors we have a distributed frame and depth buffer for the entire image.

The DDA implementation for this processor array and the data format we chose gave us an elegant, efficient, and versatile interconnection structure.

3.3 Data Structure and Algorithm Implementation

Each iteration through the inner loop of the DDA gives us the interpolated value for one pixel. It requires the divisor (dy), the quotient, the remainder, the last interpolated value, and the cumulative error (p_{sum}). Besides the cumulative error and the interpolated value, all the other values need to be calculated once for each scan line. These are calculated by interpolating along the polygon edges and are input from the polygon edge interpolator subsystem for each scan line. Thus the inner loop of the DDA constitutes scan line interpolation.

By pipelining the scan line interpolator into multiple sub-scan line interpolators we are effectively opening the inner loop of the DDA into multiple loops each with a default number of iterations equal to the extent of the sub-scan line processor, m in our case. Thus for a given scan line, y, sub-loop1 interpolates for $0 \leq x \leq m-1, y$; sub-loop2 for $m \leq x \leq 2m-1, y$ and so on. When sub-loop1 finishes, it puts all the required data into the registers for sub-loop2. When sub-loop2 starts working on these values sub-loop1 is now available to interpolate for the sub-scan line $0 \leq x \leq m-1, y$ of another convex polygon. Note that although the sub-loops are working concurrently, they are interpolating for different polygons. *For a given polygon and scan-line there is a definite order of sub-loop operations and a unique data packet in the pipeline.* Parts of this data packet, such as the last interpolated value and the cumulative error get modified as the data packet moves through the sub-loop pipeline. This imposes a simple communication structure: sub-loop$n$ communicates with sub-loop$(n+1)$ to whom it sends the modified data packet, and with sub-loop$(n-1)$ from whom it receives data packets.

If we break the loop as mentioned above, in addition to the values above the data packet needs to carry x_{start} and x_{end}, the start and end values for the loop. ($x_{\text{end}} - x_{\text{start}}$ = the divisor). Note that all the sub-loops perform the same function. For sub-loop$q$ (which interpolates for $((q-1)m \leq x \leq qm-1, y)$ this would be:

if new data in input registers and $x_{\text{start}} \leq qm-1$ then interpolate for the scan line section that lies in its domain *i.e.* from $max((q-1)m, x_{\text{start}}), y$ until $min(qm-1, x_{\text{end}}), y$; if we have not reached the last pixel of the polygon on that scan line, fill the registers of sub-loop$(q+1)$ when it is ready to accept the modified data packet.

In spite of their common functionality, the sub-loops need to know their index, q, to make the comparison, $(q-1)m \leq x \leq qm-1$, and are not identical. This implies that we need $ceil(X/m)$ different sub-loops (processors) for a screen size X pixels wide. This translates to extra logic and initialization overheads to simulate different sub-loops from identical implementations of the processor.

With a slight modification in the data format and functionality this limitation can be overcome. In the format above, the values of x_{start} and x_{end} were fixed with respect to the image axis and unchanged for all of the sub-loops. But the subloops do not need the absolute values of x_{start} and x_{end}; they need these values only with respect to their domain. For example subloop$q$ only needs to know that it should start interpolating from the third pixel in its domain. The connectivity information tells us that this "third" pixel of the subloop corresponds to $x = (q-1)m+3$. Thus if x_{start} and x_{end} could somehow be "normalized" with respect to the sub-loop indices, then the index information would be redundant and the sub-loops could be identical.

Conceptually, we normalize by associating a y-axis with each data packet. As it moves through the pipeline, the data packet shifts its y-axis. Thus the y-axis for a data packet shifts to the right by the sub-loop count, m, as the packet moves between successive sub-loops. As Figure 3 shows, by this data manipulation we are effectively moving the sub-loops to the left and keeping the modified data packet stationary. Now all the sub-loops believe that they are the starting sub-loop and hence are identical.

We implement the normalization by :

$$\begin{aligned} x_{\text{start}}(\text{to be sent}) &:= x_{\text{start}}(\text{received}) - m(\text{the loop count}) \\ x_{\text{end}}(\text{to be sent}) &:= x_{\text{end}}(\text{received}) - m(\text{the loop count}) \end{aligned}$$

$dx = x_{\text{end}} - x_{\text{start}}$, remains unchanged.

Irrespective of index, all sub-loops now perform the function:
if new data in input registers and $x_{\text{start}} \leq m-1$ then interpolate for the scan line section that lies in its domain *i.e.* from $max(0, x_{\text{start}}), y$ until $min(m-1, x_{\text{end}}), y$; if the last pixel of the polygon on that scan line has been interpolated then do not send any data packet down the scan line. Otherwise, renormalize x_{start} and x_{end} for the next sub-loop; wait until the next sub-loop is ready to accept data; fill the registers of the next sub-loop.

With this normalization, the sub-loops are all identical and the index information is hardwired in the communication structure. Thus we can replicate this one generic sub-loop for displays of arbitrary size.

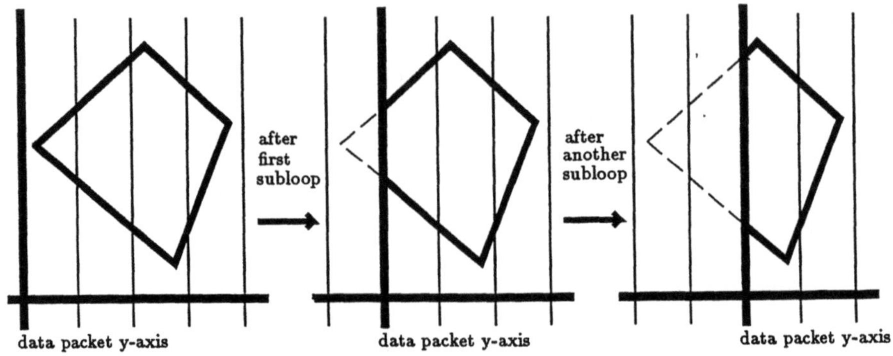

Figure 3: The effect of "Normalizing" a data packet

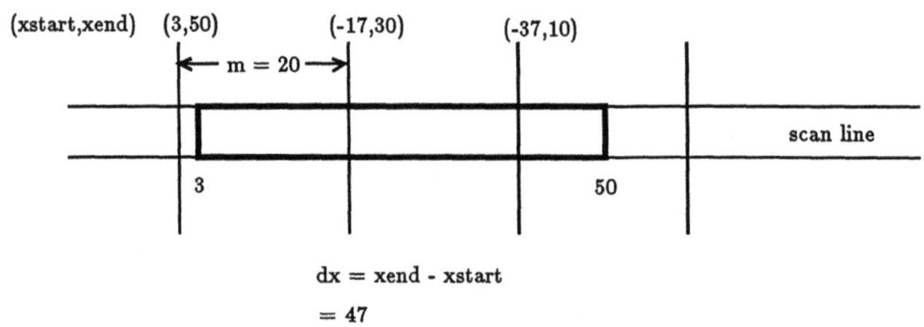

Figure 4: A numerical example to illustrate "Normalization"

3.4 Translation into VLSI

3.4.1 The Two-Dimensional Chip Array

In translating into VLSI, we mapped n of these sub-scan line processors each with its private memory onto one chip. Thus each chip is responsible for a unique m by n pixel subspace. The mapping is illustrated in Figure 5.

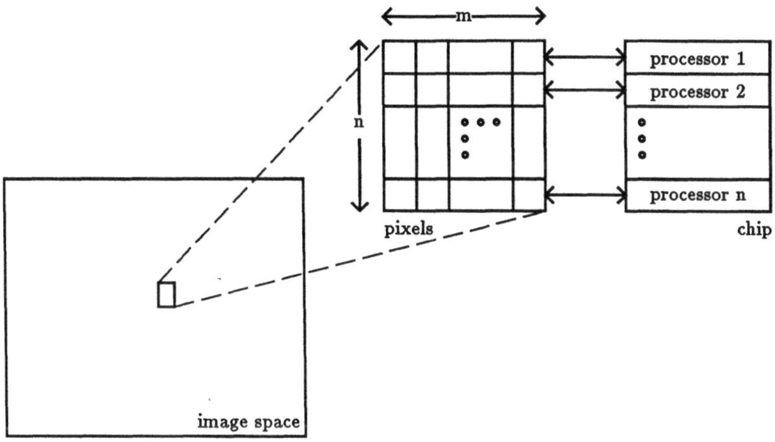

Figure 5: Processor-Pixel Mapping

All processors $i, 0 \leq i \leq n-1$, correspond to the same sub-loop index but for different loops (scan lines). Processor $i, 1 \leq i \leq n$, in chip(j,k) interpolates for pixels that lie in its domain, $((j-1)m \leq x \leq jm-1, (k-1)n+i)$. Processor i in chip$(j+1,k)$ interpolates for $(jm \leq x \leq (j+1)m-1, (k-1)n+i)$ and is the adjacent subloop to processor i in chip(j,k). Thus data packets (unique to a scan line of a polygon) move from left to right in the row of chips and only the data packets for scan lines $(k-1)n \leq y \leq kn-1$ need to be input to row k starting at chip$(1,k)$.

Subpixelling is a way of reducing the aliasing effects inherent in the depth buffer algorithm. Each physical pixel is now fractured into multiple (usually a square grid) virtual sub-pixels. We interpolate in this virtual image space that is larger than the physical image space (screen space) and use an averaging filter to merge the sub-pixels when we paint a physical pixel.

For a display size of Q by R and a subpixelling size of q by r, the array of chips would have indices $1 \leq x \leq ceil(Qq/m) = X, 1 \leq y \leq ceil(Rr/n) = Y$. Each chip communicates with the two chips on either side of it in the row, *i.e.* chip$(a,b), 1 < a < X; 1 \leq b \leq Y$, communicates with chips chip$(a-1,b)$ and chip$(a+1,b)$. Chip$(1,b), 1 \leq b \leq Y$, communicates with chip$(2,b)$ on the right and with the polygon edge interpolator on the left and chip$(X,b), 1 \leq b \leq Y$, communicates only with chip$(X-1,b)$ on the left.

If the incoming data packet is for a busy processor (with all its input buffers full) then the chip passes on a busy signal to the chip on its left so that it may refrain from overwriting this data. This communication ensures that no information is lost when we have a choked pipeline. If the data is for a sub-scan line that intersects the chip subspace, then the data packet is written into the scratch registers of the corresponding processor and the chip is free to accept more data. If the first pixel, x_{start}, lies beyond the chip domain, the x_{start} and x_{end} values are normalized and the data packet is passed onto the next chip (on the right) when it becomes free to accept this data. A normalized data packet is also passed onto the next chip when an on-chip processor completes interpolating and the last pixel on the scan line, x_{end}, does not lie in its domain. If the last pixel has been interpolated then we have finished interpolating that scan line and the data packet is not passed on. Since the polygons are clipped to the image space, it follows that chip(X, y),$1 \leq y \leq Y$, need only communicate with chip$(X-1, y)$ on the left.

Communications to the left are to indicate the busy status of the pipeline and those to the right are for transmitting scan data; this data is always normalized. The interprocessor communication is by asynchronous, source-initiated handshaking as data transfer is not lock-step with every clock tick. The communication paths for a chip array are depicted in Figure 6.

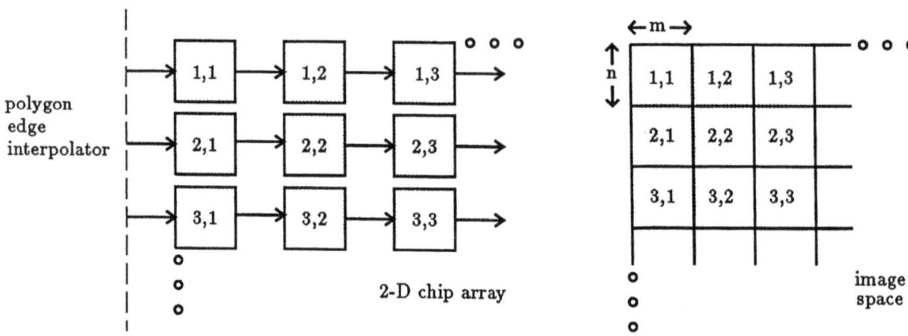

Figure 6: The communication structure of the chip array

We have a wave of data packets moving from left to right through this chip array for each polygon. The wavefront for a polygon initiates interpolation for that polygon in all the incident processors. Polygon waves may be rhythmically pumped into the system in a systolic manner.

3.4.2 Chip Architecture

Each chip has to handle handshaking and data communication for its n processors. The information for communicating with the chip to the left is distinct and independent of that with the right. In the former case we accept data packets and specify our status. In the latter, we send data packets and query the status of the next chip. Still another distinct source of data communication is with the external video circuitry for displaying the pixel values for the m by n pixels of this chip. This requires that we recognize a video-scanning signal and in response output the values of the current pixel as we cycle through all the m by n pixels.

The four distinct functional components are:

- n processors each with its private memory for m unique pixels.

Each processor interpolates for the m pixels in its domain and the memory is part of the distributed pixel and depth buffers.

- Preprocessor
  The preprocessor inputs data packets and outputs its status to the chip at the left or to the polygon edge interpolator in the case of the first chip of the row. If the data packet is not for any of the chip's processors (the starting pixel lies beyond the chip domain) it passes the packet to the postprocessor.

- Postprocessor
  The postprocessor takes data packets from the n processors and the preprocessor, normalizes them, and outputs them to the next chip. It is responsible for all the handshaking signals for this communication with the preprocessor of the next chip.

- Video-processor
  The video-processor cycles through the on-chip buffer and outputs the current pixel information in response to a video-out signal. Its access of the memory banks is transparent to, and independent of, the processors.

Additionally, each chip has a clock and timing signal generator that converts the single phase clock input to an internal two-phase non-overlapping clock. The clock generator is also responsible for producing the control signals for the preprocessor and the postprocessor and the control signals that are common to all the processor and memory banks.

This functional modularity drives the chip architecture shown in Figure 7. In the interests of brevity, only the details of the memory design are included in this document. The interested reader is referred to [Gupta 87].

The Memory Design

The total chip memory, corresponding to the pixel and depth buffers for m by n pixels, is divided into n banks, one for each processor. The memory for each processor is independent of the other processor memories.

The processor access of the pixel memory is pseudo-random; between the first and last pixels to be interpolated the pixel memory is accessed sequentially. Thus only the first pixel access is random. For the subsequent pixels, the address for the pixel just computed can be incremented to access the next pixel to be computed. We implemented this by a token that successively passes through the m pixels of the processor in the spatial order of the pixels in image-space. When the preprocessor routes the incoming data packet to the relevant processor, the "x_{start} control" unit of the preprocessor initiates the start of this token for the processor so that by the time the processor is ready to access pixel memory, the first pixel to be interpolated has the token and can send its data to the processor on a pre-charged bus. The n tokens on the chip, one for each of the n processors and its memory bank, is independent of the video token that crosses memory bank boundaries.

In the other dimension the $r+g+b+z$ bits of each pixel memory are cyclic because the processor is bit-serial and needs only a bit of information on each system clock cycle. The implementation approximates a circular shift register for the $r+g+b+z$ bits of each pixel. Memory read-out is thus destructive.

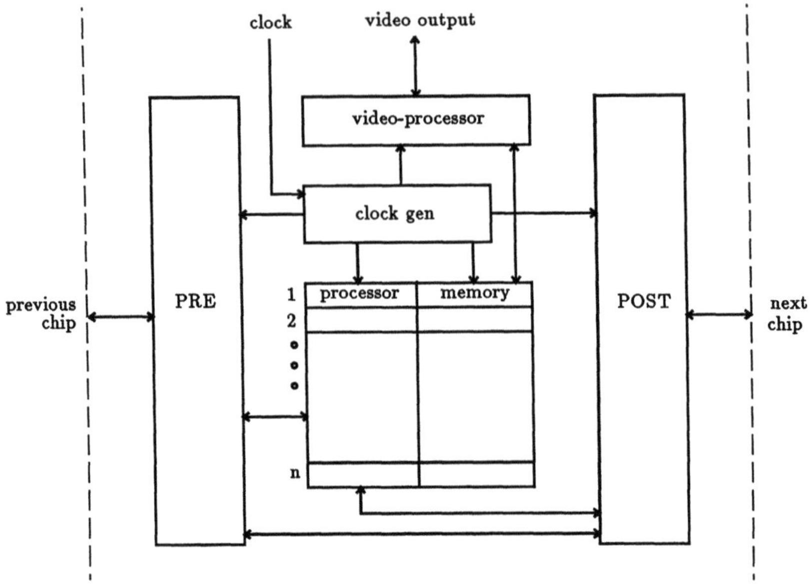

Figure 7: Chip Architecture

Figure 8 illustrates the memory organization.

4 Salient Features of the PS Architecture

4.1 Bit-Serial Arithmetic

Our architecture is communication limited. Bit-serial arithmetic helps reduce the clock period and thus boosts the communication bandwidth by a factor close to the number of bits of alternate parallel hardware. Serializing memory and logic in the design of a module while increasing the number of parallel modules improves hardware utilization and throughput at the cost of an increase in latency.

Bit-serial processors improve the flexibility of data formats. In our case, altering the data format or resolution implies minor changes in the clock and timing signal generator. It also helps in lowering the pin count and silicon area of the chip. Enhancing the connectivity of the structure, say to a hexagonal connectivity for dithering, now will not increase the pin count inordinately.

4.2 Systolic Adaptation

We have the strictly local and regular communication structure of systolic architectures. The DDA we implement, closely maps the pipelining inherent in the systolic approach. Our adaptation diverges from a purely systolic computation in that every processor does not compute on every clock tick. We use the systolic paradigm to capture the concepts of parallelism, pipelining, and interconnection structure but we do not subscribe to the strict lock-step computation of systoles. Just as in systolic systems two communicat-

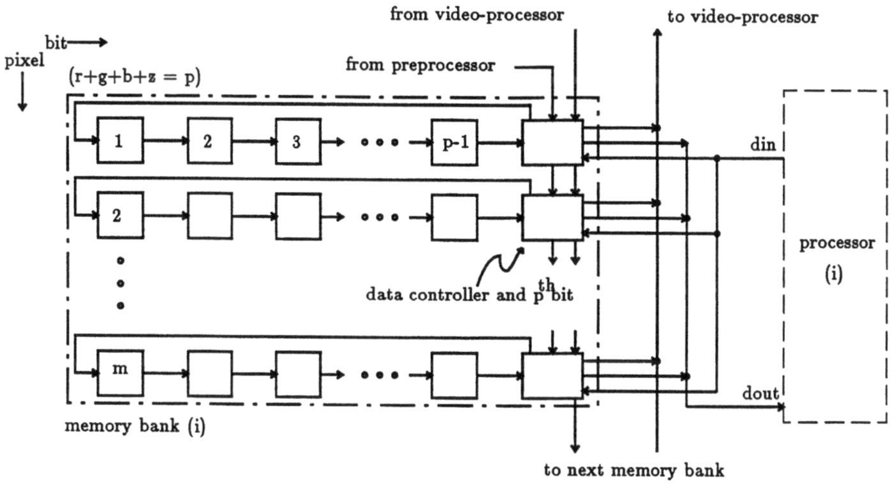

Figure 8: Memory Bank Design

ing processors have a data path between them but in our case the communication is asynchronous.

Another divergence is that the sequencing of operations is neither built into the nodes nor are control signals broadcast into the array in an SIMD fashion; in our adaptation the control for the computational array is distributed into the array. In the current implementation, the data encodes the operation – normalize, interpolate for a variable number of pixels, kill the data packet – mimicking a data-flow architecture. In manipulating the data packet each processor is effectively encoding the command for the next processor. Additionally, the data packet also contains an explicit command field. Currently the only encoded command controls z-compare. PS allows each processor to affect the operation of subsequent processors in the scan line by manipulating the data and also by explicitly manipulating the command field. We hope to exploit this flexibility to address complex applications.

4.3 On-chip Memory

Scan line conversion requires a high memory bandwidth. Keeping in mind the penalties that off-chip accesses entail in terms of power, speed, pin count, and silicon area, we have a compelling reason for on-chip memory. There are other not so obvious advantages of on-chip memory. Distributed memories that are in close proximity to the processor offer less memory contention and global address decoding. The on-chip memory and processor share a common clock reducing concerns of clock skew.

We can exploit the VLSI advantage by specializing the on-chip memory to the processing task. In our case this corresponded to a multiple-ported, dynamic memory that mimics a circular shift register for each pixel. It is optimized for bit-serial arithmetic. The data dependencies and locality that are typical to our processing task give us a simplified, fast, and small memory decoder, and reduced wiring.

4.4 Algorithm Implementation

The algorithm implementation for the distributed system tracks the simple and local connectivity of the systolic structure. The data structure carries all relevant information thus avoiding the need for common busses or global communication. It automatically tesselates polygons onto the underlying chip array structure adding extensibility to the implementation. Tesselation does not incur any space or time overhead as would be the case in a higher level partitioning scheme.

4.5 Hierarchical and Distributed Control

The primary considerations in the design of the control circuit are the silicon area, the speed of the chip, and the design time. For VLSI these are direct manifestations of the cost of the circuit. We employed a hierarchical design that separates the sequencing and the generation of commands [Obrebska].

In our design the clock and timing signal generator forms the top level of the hierarchy of the control. By extracting the timing signals that are common to the different functional units of the chip we avoid repetition and redundancy in the signal generation. These signals form the skeleton of the control structure providing validation and supervision for the flesh of the structure within and between the units. This flesh is distributed at various levels. The distribution gives us flexible and localized control, and minimizes long-distance communication.

The clock and timing signal generator is implemented as a PLA. The PLA construct is easy to generate, test, and debug. Its use at this level makes the implementation flexible. We can change the parameters of the implementation *viz.* the system vs. processor clock ratio, the data format, the buffer sizes, the screen resolution, the number of pixels per processor (m), with minor modifications. The versatility of the PLA structure, the irregular nature of the control logic, the large size of the corresponding FSM, and the fact that there is only one of these per chip justify the area of the PLA.

Token-passing is used to control communications between units. Pre, post and processor communications use one polling token, video communication employs a token that circulates through all the m by n pixels of the chip, and each processor-memory bank communication is controlled by an independent token; thus each chip has $n+2$ independent tokens. All information transfers from the processors to the postprocessor, from the preprocessor to the processors, and between each processor and its pixel memory bank are initiated by the processor clock. A processor clock period being $r+g+b+z$ times the system clock period, we have $r+g+b+z$ system clock cycles for decoding communication requests. At the same time, we want to avoid personalized control lines for every type of communication. Token-passing gives us the best tradeoff between decoding time, size of decoding circuitry, and multiplicity of control lines. The use of a shifting token for processor-memory bank control exploits pixel coherence (interpolation for pixel $i+1$ follows that for pixel i). The video-memory token requires no decoding.

Within each unit, control is implemented by gates and dynamic shift registers mimicking the "delay-element method" [Hayes]. This design style conserves area and reduces delays because of the small incremental cost of latches in MOS technologies. The internal functions of the processors are directly controlled at this level. It is therefore repeated as many times as the number of processors per chip (n); this makes the area considerations significant. The regularity of the algorithm and the "tight" properties of this design style justify its non-uniformity and thus lack of automation.

4.6 Logic Implementation

The use of steering logic allowed us to exploit the features typical of the implementation technology, namely CMOS. We have pass transistor logic networks with the logic level appropriately restored. These networks control data or signal flow between data latches – another elegant and low cost feature of MOS logic.

Static CMOS logic, however, requires us to implement both the complimented and uncomplimented forms of the function in n-type and p-type transistors respectively. Not only is silicon area wasted in the duplication, but the problem of routing signals is aggravated. We used precharge logic whenever possible. Besides the savings in circuitry and routing, precharge logic gives us lower capacitances and hence higher speeds. With only one p-type transistor in the pull-up path, load capacitances charge faster. We easily derived mutually exclusive compute-precharge phases from the two-phase non-overlapping clock on the chip thus fulfilling the precharge requisite that inputs to precharged busses be glitch-free and that the bus not be used for the time it needs to precharge. We opted not to concern ourself with the lower speed bound in using dynamic logic.

5 Comparison with Existing Polygon Rendering Systems

Our design exploits parallelism by partitioning the image space so that each processor is responsible for only a sub-image and consequently fewer objects. Another possible approach is to partition the object space so that each processor is responsible for only a subset of the objects in the image. We compare the PS architecture with four prominent and recent architectures for scan conversion that apply image space partitioning. These are:

- Pixel-planes by Henry Fuchs et al. at the University of North Carolina [Fuchs 81,Fuchs 83,Poulton et al. 85]
  Pixel-planes is a processor-per-pixel architecture for Gouraud shading and depth-buffering polygons. The designers have since discovered other interesting applications that map onto the functional paradigm of the architecture. The most unique aspect of the architecture is the use of multiplier trees to calculate the linear equation, $F(x) = Ax + By + C$, at each pixel. Differently said, the multiplier trees decode linear equations onto the pixel space.

- Scan Line Access Memories by Stefan Demetrescu at Stanford University [Demetrescu 85]
  Scan Line Access Memories, or SLAMs, use conventional high density RAMs enhanced with limited processing for high speed rasterizing of large sections of scan lines. The current implementation runlength encodes a specified 16-bit halftone pattern to affect an entire, or a subset of, a scan line. The architecture draws strength from its simplicity and the use of conventional RAM design concepts.

- Super-Buffers by Nader Gharachorloo et al. at Cornell University [Gharachorloo 85]
  The Super-Buffer architecture is a systolic approach to the task of rasterizing polygons. Super-Buffers, developed independently, share many features with PS. Notably, we both divide the task of rendering polygons into polygon edge interpolation and scan line interpolation. In polygon edge interpolation both systems employ an incremental algorithm. In addressing the communication problem, our approach is a variation of the systole to which the Super-Buffers adhere strictly.

- Systolic Array Graphics Engine by Nader Gharachorloo et al. at IBM, Thomas J. Watson Research Center [Gharachorloo et al. 88]
  Systolic Array Graphics Engine, or SAGE, is an update to the Super-Buffer system. Also developed independently of PS, the SAGE architecture is very similar to Super-Buffers with the added functionality of Gouraud shading. The primary differences between SAGE and PS remain the processor-per-pixel of SAGE versus the processor-per-subscanline of PS, the external frame–buffer requirement of SAGE versus the distributed frame–buffer of PS, and the common video and processing clocks of SAGE versus the independent video and processing clocks of PS.

In our comparison, we characterize the efficiency of an architecture by:

- the extensibility of the architecture to increasing functionality.
- the extent of host involvement.
- the speed potential.
- the processor utilization.
- the communication bandwidth.

Only PS, Pixel-Planes, and SAGE currently implement Gouraud shading and depth-buffering. SLAMs and Super-Buffers have a one-bit plane at each pixel and do not support hidden surface removal.

Pixel-Planes inherently support all functions that can be represented as a linear equation. Functions that do not have the linear equation as their natural representation require extensive preprocessing. For example, polygons are most conveniently described in terms of their vertices. Edge equations are easily obtained from this information; however planar equations for color require more preprocessing. Super-Buffers and SAGE are limited in that no memory is retained past scan line boundaries, requiring all pixel information to be retransmitted through the row of Graphics Engines at the onset of scan lines. This requires pixel information for the entire image space to be stored externally along with data for all the active polygons. The SLAM architecture is optimally designed for a 16-bit function repeatedly applied to all pixels on a scan line. The architecture loses most of its elegance and performance metrics for more involved processing. The PS architecture is optimized for incremental arithmetic and local communication. With the understanding that all global communication can be implemented as local communication using systolic conversions [Leiserson a,Leiserson b], we foresee few limitations to applications with our architecture.

Global communication in Pixel-Planes and SLAMs limit the performance potential of the systems. In Super-Buffers and SAGE, the image clock is also the video clock, limiting the polygon throughput. In all of these systems the memory is essentially single-ported; thus, video access and image processing are not completely transparent. Our simple decoding structure gives us multi-ported on-chip memory with completely independent and transparent video and image accesses. Also, in all the other architectures that we have considered all the chips are functionally different. This translates to additional logic, delay, or host involvement in an initializing phase. With a simple "normalizing" function, we avoided that requirement.

Processor utilization in Pixel-Planes is low by virtue of the processor-per-pixel approach to a lock-step-with-polygon character enforced by the global communication. In SLAMs, processor utilization is high only for applications with a repeatedly applied pattern covering large sections of a scan line. Super-Buffers, SAGE, and PS enjoy a high processor utilization profile. In Super-Buffers and SAGE, however, the processor-per-pixel architecture witnesses many processors simply performing the x_{left} and x_{right} comparison. This dilutes the effective processor utilization. In the same token, the processor-per-subscanline of PS might witness bottlenecks when consecutive data packets are destined for the same processor. In the worst bottleneck case, the enhanced processor-pixel mapping of Subsection 6.1 loses a factor of m, where each processor is responsible for an m-pixel sub-scanline, in comparison to Super-Buffers or SAGE. This underlines the tradeoff between processor utilization and speed potential, or the average case versus the worst case.

Of all the architectures reviewed in this comparison, the SAGE architecture bears the closest resemblance to PS and merits a closer inspection. In SAGE just as in PS, the boundary between the polygon edge interpolator and the scan line interpolator limits the communication bandwidth of data packets along a scan line. This observation in conjunction with the processor-per-pixel approach justifies the purely systolic communication of SAGE. In the processor-per-subscanline or processor-per-multiple pixels architecture of PS, however, the number of pixels interpolated by a processor may vary from 0, in cases of sub-scanlines that do not intersect the domain of the processor, to m, in cases of sub-scanlines that overlap the entire m-pixel domain of the processor. Thus the processing time per data packet is not constant, necessitating the asynchronous hand-shaking communication in PS.

The absence in SAGE of a distributed frame–buffer is closely tied to the reuse of the processing clock as the video clock; this is by far the most significant difference between SAGE and PS. When rendering more complex scenes than those permitted by the video rate, SAGE requires an external frame–buffer for storing the output of the SAGE chips. This implies either that the buffer values for a scan line have to be shifted into the vector of processors in time before the new interpolation for the same scan line, or external circuitry is needed for correlating the new and previous values at each pixel. The distributed frame–buffer and independent video and processing clocks of PS permit graceful video degradation with fewer frames per second. Even in the case of simpler scenes, the processing rate matches the video rate only for four interleaved SAGE chips. SAGE appears to have no way of recovering and retransmitting the error term of the interpolation thus potentially introducing aliasing artifacts at the boundary between the chips.

The extent of host involvement is significantly different between the SAGE/Super-Buffer and PS architectures. Consider the following abstraction: the task of depth-buffering is effectively the sorting of all polygons along x, along y, and then for a given (x, y), *ie.* for each pixel on the display, the sorting of all relevant polygons along z. For each pixel, the z-buffer retains the intensity data only for the closest polygon. The sorting along x and y is simply an optimization to avoid processing all polygons at every pixel. Polygon sorting along x is circumvented by providing a processor at each pixel along the scan line; the z-sorting of all polygons for all values of x is then off-loaded from the host to PS/SAGE/Super-Buffers/etc. Similarly, scan line processors for each scan line in PS circumvent the need for polygon sorting along y; the z-sorting of all polygons for all values of (x, y) is off-loaded from the host to PS. SAGE/Super-Buffers on the other hand, only z-sort polygons along x, the sorting of polygons along y continues to remain the responsibility of the host. In this sense PS is an evolution of the SAGE/Super-Buffer architectures.

6 Application Possibilities and Enhancements of the PS Architecture

This chapter is organized in an extensibility precedence: the extensions to the system are described in a decreasing order of detail and/or an increasing magnitude of modifications. All the enhancements share the same design philosophy – on-chip memory, decoding that exploits data dependencies, local communication, bit-serial processing – as the current system, and most cases propose a simple increase in the functionality of the processor.

6.1 A Better Processor-Pixel Distribution

The connectivity of the chip array in the current implementation introduces an asymmetry in communication. The communication requirements decrease as we move from the first column of chips to the last as the first chip in the row has to transfer/process data packets for all the chips in the row. As we move down the row, in increasing x, data packets terminate and the rate of data flow between chips falls.

Currently, the n processors of a chip are stacked vertically so that each chip is responsible for a square array of pixels. However, if we connect the processors back-to-back so that they span contiguous sections of the same scan line, we continue to have XY/N^2 chips in the system but we change the aspect ratio of the array. The chip array changes to X/N^2 by Y from X/N by Y/N, and the processor array is the same at X/N by Y. However with X/N processors per scan line and only one scan line per row, we now have X/N processors per row of chips and not X processors per row as earlier. Also, data packets span potentially larger sections of a chip's domain. These reduce the data packet requirements per row of chips. Conversely, with a chip now processing larger scan sub-sections the data traffic between chips falls. Thus we help remedy the asymmetry of communication discussed above.

The lower communication requirements at the same communication bandwidth improves throughput and processor utilization. In the context of locality of reference which has data for scan line $i+1$ following that for scan line i, we increase parallelism by assigning one scan line per row of chips. Note that we have the same number of processors per scan line leaving the performance contribution of pipelining intact.

In this approach consecutive processors span contiguous sections of the pipeline. In the context of pixel coherence, the data dependency that applies to adjacent pixels within a scan line, we can now simplify and enhance the power of the token decoders for higher performances.

6.2 Processor Enhancements

We only consider applications that can take advantage of the system architecture. We expect that many algorithms could be reformulated to exploit the distributed control, communication and processing structure of the system. The list of possible applications is not purported to be complete. The last two in the list below embody the structure which promises interesting solutions to difficult problems.

6.2.1 Anti-aliasing

Aliasing effects arise from the finite size of pixels. Differently said, it is the quantization error of converting real pixel values to the integer precision of the display. In the DDA computation although the data packet carries only integer information, all the precision of the real number interpolation at each pixel can be reconstructed from the values in the data packet. As an example, p_{sum}/dy, the difference in the integer x pixel value we adopt and the true x value at the scan line, is available as the integer values p_{sum} and dy. We can use this information to color edge pixels so that edges appear anti-aliased. Note that this is only an approximation to true anti-aliasing and hence is not free from defects.

6.2.2 BitBlt

By BitBlt [Foley 84] here we mean copying or moving arbitrary sections of the screen to another location on the screen with one command.

The pixel data now originates from the source sub-screen and is routed to the destination sub-screen through the switch network between the polygon edge and scan line interpolator systems. A data packet with the source pixel location, length on the scan line, and destination scan line and pixel location moves through the scan line and affected pixel data is appended to the data stream at the source and stripped from the top of the stream at the destination.

6.2.3 Other Incremental Algorithms

The architecture supports algorithms that can be described incrementally. Bresenham's Circle algorithm [Foley 84] is a typical example. It ideally maps the architecture and can be implemented with only minor enhancements of the processor, making the circle throughputs of the system comparable to the polygon throughputs. Other interesting classes of algorithms and graphical effects can be obtained from the current data structure. A case in point involves successively changing the constants in the data packet. By left-shifting the quotient at each computation, we can trivially compute the function 2^x where x is the pixel attribute. Algorithms that can optimally utilize the parallelism of PS can be found in many disciplines. These algorithms run inhibitively slowly on current mainframes and the performance improvement by running them on PS would make them practical and useful.

6.3 Fault-Tolerance

We can incorporate a measure of fault-tolerance by having an extra processor and memory bank per chip and storing the bank attribute (good/bad) in the token decoder. If the processor or memory of any one of the n banks of a chip, say p, is found to be defective the token decoder delays by one clock cycle the valid signal for banks that follow the defective bank in the token path. In this simple way banks $p \ldots n-1$ are effectively replaced by banks $p+1 \ldots n$, where banks $0 \ldots n-1$ are the chip's n banks and bank$n$ is the extra bank added to tolerate a faulty bank.

The detection of a problem is simplified by the existence of the postprocessor; since all banks are identical the same scan line can be sent to successive processors and the result compared in the postprocessor. We do require some additional logic on the chip to do this and to reflect the result in the token decoder. The loss in area incurred by this fault-tolerant measure is less than $1/m$ times the chip area. Also, the chip forms a generic PS stage and we can replace a faulty chip with a new chip which will reset itself with the background polygon at the start of the next frame. We are not required to interrupt and reinitialize the entire system.

7 Conclusions

PS exhibits the characteristics of distributed systems: increased performance, extensibility, and modular architecture. An eclectic bag of architectural methodologies directly addresses the limitations on communication bandwidth and processor utilization. The power of PS lies in the extensive pipelining and parallelism it incorporates. Bit-serial arithmetic, systolic connectivity, on-chip memory, and hierarchical control effect the performance metrics and extensibility of PS. In its implementation, PS exploits the fabrication technology, CMOS.

A number of tradeoffs and marriages make the PS architecture unique: the distributed communication of systoles leading to high bandwidth and the independence of asynchronous communication; decoding time *vs* the minimal control wires as in token-passing; the low processor utilization of a processor per pixel *vs* the communication and processing bottlenecks of an m by n approach.

In the context of the current application, the reformulation of the DDA maps the architecture efficiently. A judicious processor-pixel distribution is shown to enhance the symbiotic relationship of the application and the architecture. The proposed architecture exploits the data dependencies peculiar to the task of scan conversion. The exercise underlines the advantages of concurrently developing the algorithm, data structure, and architecture.

In designing the PS system, we witnessed an anachronism in design psychology: "memory" = "RAM". Even for the on-chip memory of specialized hardware, the organization is or closely resembles a RAM. However in most cases, certainly in ours, random access is overkill; And there is a price to pay for it! For most applications, data coherencies can lead to faster, smaller, and more efficient memories. Semi-random access memories are finer tuned to the application and therefore more effective. A case in point are the *multiple ports* of the PS on-chip memory.

A Implementation Details

A.1 Architecture Details

m, the number of pixels per processor = 16
n, the number of processors per chip = 16

Pixel Data Format:
r, g, b = 8 bits each
z = 16 bits
Therefore the depth buffer stores 40 bits per pixel. This corresponds to:
40×16 = 640 bits per memory bank, and
$16 \times 40 \times 16$ = 10,240 bits of memory per chip.

1 processor cycle = 48 system clock cycles.
The chip was designed to run at clock rates of 40 Mhz and over.

A.2 System Performance Numbers

$p = 48$ clock cycles per pixel
$c = 40$ MHz
$X = 2^{10}$
$Y = 2^{10}$
$N = 16$
$v = 6$ clock cycles per pixel

the best pixel rate $= cXY/(pN) \approx 5.4613 \times 10^{10}$ pixels/sec.
the worst pixel rate $= c/p \approx 833,333$ pixels/sec.
the best polygon rate (current architecture) $= cY/(pN) \approx 5.3 \times 10^7$ polygons/sec.
the best polygon rate (modified architecture) $= cY/p \approx 8.533 \times 10^8$ polygons/sec.
the worst polygon rate $= c/(pN) \approx 52,083$ polygons/sec.
the best frame rate $= c/(vN^2) \approx 26,041$ frames/sec.
the row pixel-bus frame rate $= c/(vXN) \approx 407$ frames/sec.
the column pixel-bus frame rate $= c/(vYN) \approx 407$ frames/sec.
the worst case row pixel-bus latency (current) $= (Xp/N + p + XNv)/c$ seconds. $\approx 2.5356 \times 10^{-3}$ seconds.
the worst case row pixel-bus latency (modified) $= (Xp/N^2 + p + Xv)/c$ seconds. $\approx 1.596 \times 10^{-4}$ seconds.

A.3 Chip Details

Technology: 3 micron cmos/bulk
Silicon Vendor: MOSIS

Thus the 3-transistor per bit dynamic memory requires $16 \times 40 \times 16 \times 3 = 30,720$ transistors.

Chip size: $9mm \times 4mm$
Pin count: 18
Transistor count: 50,000

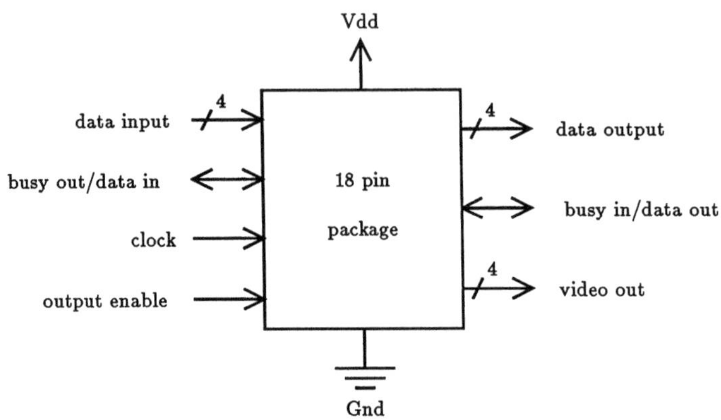

Figure 9: The PS Chip

References

[Demetrescu 85] Stefan Demetrescu. High speed image rasterization using scan line access memories. *1985 Chapel Hill Conference on VLSI*, 1985.

[Foley 84] J.D. Foley and A. vanDaam. *Fundamentals of Interactive Computer Graphics*. Addison-Wesley, 1984.

[Fuchs 81] Henry Fuchs and John Poulton. Pixel-planes: a vlsi-oriented design for a raster graphics engine. *VLSI Design*, Third Quarter 1981.

[Fuchs 83] Henry Fuchs, John Poulton, Alan Paeth, and Alan Bell. Developing pixel-planes, a smart memory-based raster graphics system. 1983.

[Gharachorloo 85] Nader Gharachorloo and Christopher Pottle. Super buffer: a systolic vlsi graphics engine for real time raster image generation. *1985 Chapel Hill Conference on VLSI*, 1985.

[Gharachorloo et al. 88] Nader Gharachorloo et al. Subnanosecond pixel rendering with million transistor chips. *SIGGRAPH '88 Conference Proceedings*, August 1988.

[Gouraud 71] H. Gouraud. Continuous shading of curved surfaces. *IEEE Transactions on Computers*, June 1971.

[Gupta 87] Rajiv Gupta. *PS: Polygon Streams, A Distributed Architecture for Incremental Computation Applied to Graphics.* Master's thesis, California Institute of Technology, 1987.

[Hayes] J.P. Hayes. *Computer Architecture and Organization.*

[Leiserson a] Charles E. Leiserson. *Area-Efficient VLSI Computation.*

[Leiserson b] Charles E. Leiserson and James B. Saxe. Optimizing synchronous systems. *Journal of VLSI and Computer Systems*, 1(1).

[Obrebska] Monika Obrebska. Comparative survey of different design methodologies for control part of microprocessors. *VLSI Systems and Computations.*

[Poulton et al. 85] John Poulton et al. Pixel-planes: building a vlsi-based graphic system. *1985 Chapel Hill Conference on VLSI*, 1985.

A Generalised Parallel Architecture for Image Based Algorithms

G. J. Vaudin, G. R. Nudd, T. J. Atherton, S. C. Clippingdale, N. D. Francis, R. M. Howarth, D. J. Kerbyson, R. A. Packwood and D. Walton

Department of Computer Science
University of Warwick
Coventry, UK

1. Introduction

Real time image generation and image understanding require levels of computing power, that are beyond that available from conventional sequential machines. Current commercially available systems aimed at this area make use of special purpose hardware to achieve the necessary throughput, but these systems can only achieve their performance for a restricted set of algorithms that are implemented in the hardware. A programmable general purpose parallel machine offers the possibility to achieve the required performance without restricting the choice of algorithm.

Unfortunately it is by no means clear which parallel architecture should be used. Many general purpose parallel architectures have been proposed but none has proved universally applicable, their problem being that their performance tends to be highly dependent on the algorithms that are being used, and it is therefore difficult to claim any of them are truly general purpose. However parallel machines can still be highly effective in specific problem areas where the class of algorithm is known.

Our aim has been to design a parallel machine that is optimised for image based algorithms in both graphics and image understanding. The architecture is not limited to a specific set of algorithms, but is instead optimised towards a class of algorithms which we believe are representative of image based algorithms. This has not been a paper study, but has resulted in us implementing such an architecture. We have achieved this by making use of industry standard components and integrating them into a system level architectural design. Also we have where possible used industry standard programming languages to program our machine.

2. Existing Parallel Architectures

Various classifications exist for parallel machines. The most fundamental is Flynn's [5] which classifies machines by the number of instruction streams that they employ. Single instruction / multiple data (SIMD) machines have a single instruction stream which is common to all the processors in the system such that each instruction operates on many different data in parallel. Multiple instruction / multiple data (MIMD) machines have a separate instruction stream for each processor so that many different data may be acted on independently from one another.

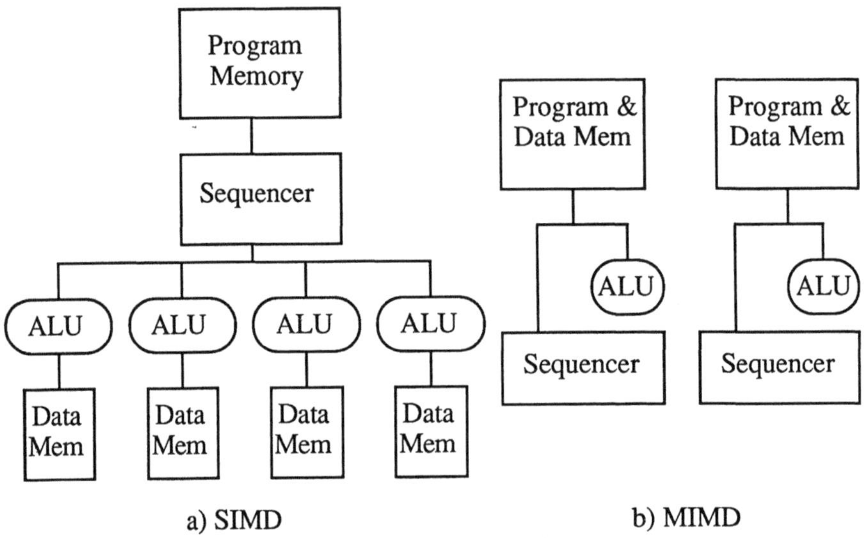

Figure 1: SIMD and MIMD Architecures

SIMD machines [1,2,3,4] excel at problems where large arrays of data are to be processed identically. They are also very good at communication intensive problems, provided that the communication is also identical for each processor. Since SIMD machines have only a single instruction stream they require only one program memory, and one instruction sequencer. This means that more of the machine can be given over to computational units and so makes them much more cost effective than MIMD architectures if the data is to processed identically. However SIMD machines become progressively more inefficient the more the processing begins to vary from one processor to another. Most SIMD architectures rely on being able to exclude certain processors from a particular piece of computation to implement different processing on different processors. This leads to processors lying idle for some of the time, and the more the processing varies the more time processors spend idle. The limiting case being when each processor is executing a completely separate instruction stream at which point the performance degrades to that of a sequential machine.

Conversely MIMD machines [13] excels at problems where each processor is completely independent of each other. Since it has instruction memory and instruction sequencing on every processor there is no overhead involved in every processor executing a different instruction. However because the processors have their own instruction streams there is no global synchronisation between processors, which means that communication between processors involves some form of rendezvous, which implies that one or other processor has to wait for the other. This leads to an increased communications overhead compared with the SIMD machines. Also it is characteristic of algorithms that run well on MIMD machines that because of the independent nature of the processing on each processor the communication tends to be less localised than with algorithms that run well on SIMD processors. This further increases the communications overhead, and makes MIMD machines less suitable for communication intensive problems.

In addition to the MIMD/SIMD classification parallel machines can also be classified into coarse and fine grained. Coarse grained implies that the machine is made up from relatively large processors, usually with floating point hardware, whereas fine grain implies the use of much smaller processors many of which can be integrated onto a single chip. Coarse grained machines are generally good at general purpose "scientific" computations which make heavy use of floating point calculations, and which therefore benefit from having these calculations done in hardware. Fine grained machines are generally better at tasks which involve fairly simple integer arithmetic often involving variable word lengths. For these tasks the use of a large arithmetic unit would be very inefficient since most of it would lie idle, a better use of the available resources is to implement more of the simpler processors.

In general most SIMD machines tend to be fine grained, whereas most MIMD machines tend to be coarse grained. It would be very inefficient to implement a fine grained MIMD machine since the overhead of providing an instruction sequencer for each processor would begin to dominate the system. There is no such restrictions with SIMD machines so it is reasonable to build systems with large numbers of small processors. It has been argued that it is in general more efficient to use a large number of simple processors than a smaller number of more complex ones [14], but as I have already mentioned this is in fact only true for a restricted class of problems. There may therefore be applications for which a coarse grained SIMD machine would be ideal, but at this time most available SIMD machines are fine grained.

3. Image Based Algorithms

To decide which of these architectures would be most suitable for the use in image based applications we need to look at the structure and composition of the problems and the kind of algorithms that are used in them.

3.1 Image Understanding

An image understanding system is required to take an image (typically consisting of an array of intensity values) as input and produce a high-level abstract description of the scene as output, in real time. To do this it would proceed in a number of stages, at each one refining the data into a more abstract form.

Initially low-level or iconic-to-iconic processing would be performed, which processes the input image and produces a modified form of the image as output (this is the image processing part of the problem). The result of this stage of processing will be a segmented and labelled image, where the various areas of interest within the image would have been isolated and uniquely labelled. These areas may be edges, enclosed regions or other more specific features.

The next level of processing consists of iconic-to-numeric processing where quantitative information about the isolated regions is extracted from the image. This information might include the centroid, area and variance of a particular region for example. The information required would depend on the type of feature being examined.

Finally numeric-to-symbolic processing is performed which takes this extracted data and uses it to form hypotheses about the scene represented by the image. This would involve using a priori knowledge of the possible contents of the scene to identify the relevant features and then match them with hypothesised objects. The system can then perform reasoning based on the interrelationship of these hypothesised objects in order to produce a full description of the scene.

One of the difficulties in building a machine to do this processing is that a wide variety of operations need to be performed, on widely differing representations of the data. An architecture which is good at one type of operation (eg. pixel-based iconic-to-iconic) is likely to be unsuited to other types (eg. symbolic).

For the pixel or iconic representation, and the corresponding class of operations such as thresholding, convolutions, and histogramming that one wishes to perform on this data a very high throughput is necessary to process the large amount of data. About 10 million pixels need to be processed per second for real time applications, and the throughput is particularly large when one considers that even for a simple linear operation such as a 5x5 convolution 25 multiplications and accumulates are required per pixel. For non-linear operations such as median filtering the throughput grows as N^2, and of the order of 1,000 MOPS may be needed.

To meet this throughput requirement requires a very high degree of parallelism, ideally one processor for every pixel in the image. However it should be noted that the operations are carried out globally on all pixels simultaneously and are therefore suitable for implementation on an SIMD machine. Also the operations tend to be numerically simple, with 8-bit integer multiply probably the most complex operation required. Thus an array of simple processors could be used which would be sufficiently small as to make such a massively parallel scheme feasible. Many such machines, mostly consisting of simple bit-serial broadcast SIMD array processors, have been built (CLIP, DAP, MPP [1,2,3,4] etc.) and have proved very able at this low level image processing type of problem.

The iconic-to-numeric stage, is responsible for analysing the results produced by the iconic-to-iconic phase. It needs the ability to extract results from the pixel array, allowing for the fact that the results required may vary depending on the feature that is being processed, and to process these and pass them on to the symbolic layer. The data extracted will be more complex in nature than the simple intensity values of the iconic stage, but will be less numerous, and the processing performed on it will typically be more varied than is the case with iconic data, since the data itself is more varied.

To reflect these needs a suitable processor would be a medium grain MIMD machine which would be able to influence the processing of the iconic data which it is to use as input. A conventional microprocessor would be a possible candidate for this purpose, with some additional hardware to allow it to communicate with and have some control over the iconic processors from which it receives its input.

The symbolic processing requirements are more varied in nature. By this stage any direct relationship with the image data is gone, and the processing deals with more abstract notions of shape and relationship. These structures will be be much more complex than in previous stages but will be correspondingly less numerous. For this purpose a powerful general purpose MIMD parallel machine would appear to provide the appropriate choice.

The conclusion to be drawn from this is that no one architecture will be suitable for all stages in the image understanding problem. The ideal image understanding architecture would consist of a variety of different processors arranged in such a way as to allow data to flow between the different sections.

3.2 Computer Graphics

Computer graphics can be regarded as the reverse process to image understanding, instead of starting with an image and producing a model, one starts with a model and produce an image which represents that model. Like image understanding, computer graphics systems proceed in a number of stages, in this case moving from an abstract representation of a scene towards a concrete pixel array representation.

The first step in processing consists of viewing transformations, where the model is transformed into the coordinate system of the viewer. This may involve different transformations for different objects in the scene, if the relationship between the various objects is not fixed. Working out these relationships may be a complex problem in itself if, for example, the different object were aircraft in a flight simulator. Then the illumination of the objects is calculated based on their positions with respect to the light sources. Most systems will perform some clipping at this stage, where objects out of view are removed from further calculations.

Next the objects are decomposed into their constituent parts, usually planar polygonal patches, and these are then projected onto the two dimensional screen coordinate system. Further clipping is carried out to remove parts of objects which fall outside the boundaries of the screen, and surfaces which are obscured by other objects may be removed. Finally the pixels which represent the 2D polygonal patches are evaluated, and each set to the appropriate intensity level. This stage may attempt anti-aliasing, and smoothing and may also be responsible for hidden surface elimination.

The first stage in the process involves a large number of floating point operations to be performed on the objects, which have no fixed interrelationship, and may have different representations and thus may require different processing. These operations can in general be performed entirely independently of each other, so that inter processor communications is not generally a problem. An MIMD array of floating point processors would appear to provide an appropriate solution to this stage.

Once the data has been converted to screen polygons the processing becomes arithmetically simpler, usually involving integer arithmetic only, and also tied more closely to the physical layout of the screen memory. This kind of problem is amenable to solution by an SIMD array of integer only processors. However there will typically be a many to one relationship between polygons and the pixels which make up the polygons, which implies that there will be fewer polygon processors than pixel processor, which in turn implies that a more coarse grained processor is used.

The final pixel tiling involves a very large number of simple operations to be performed identically over the pixels forming a given polygon, and so a large SIMD array of simple processors would be appropriate.

Similarly to image understanding it can be argued that no one of the available architectures is ideally suited to all the stages in the graphics process. What is required is a machine which consists of a variety of different processors arranged in such a way that data can be efficiently transferred between them.

We have designed and constructed such an architecture which provides four distinct "layers" of processors. These processors can communicate both within one layer or between adjacent layers. In this way data can be passed from one stage to the next. In the following sections we will describe in detail the design and implementation of this architecture, and then go on to describe the programming of it.

4. Warwick Pyramid Machine

The Warwick Pyramid Machine [6], has been developed to address the needs of a specific class of image based problems, namely image understanding and computer graphics. It consists of four distinct layers of processors, each optimised for one stage in the problem. Since data flows both within each layer and between adjacent layers, it is convenient to regard the layers as being stacked one above the other. Also as each layer contains fewer processors (but more powerful ones) than the layer below it, we regard the machine as being a pyramid of processors.

At the base of the pyramid is the iconic layer, which is a fine grained Multi-SIMD machine. This consists of an array of independently controlled SIMD machines called clusters. Each cluster contains its own controller which provides it with its instruction stream. These controllers are each associated with a symbolic processor which processes the results from the cluster and also determines the instructions generated by the controller. The symbolic layer is an array of coarse grained MIMD processors. The symbolic layer is connected to a conventional host machine which is used to interact with the pyramid.

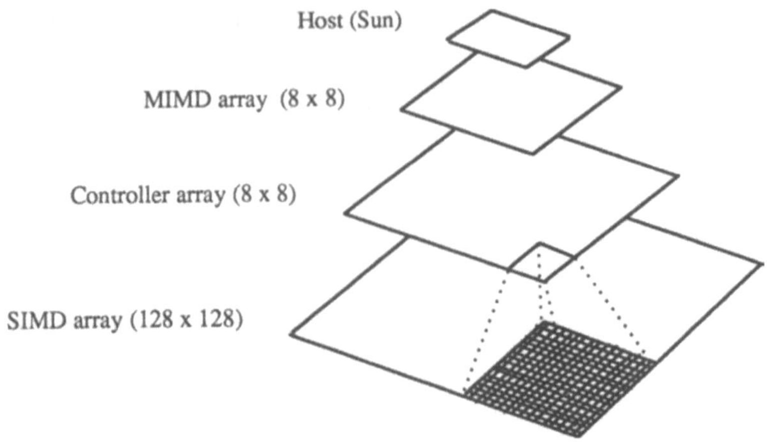

Figure 2: The Warwick Pyramid Machine

4.1 Host Machine

At the apex of the pyramid is the host computer, which performs no computations of its own but provides the user interface for the pyramid below it. The host machine provides mass storage and is responsible for downloading programs into each processor. It also provides a familiar user environment, such as editors and debuggers for developing programs to be run on the lower levels, displaying intermediate results, and so on. A graphics workstation such as a SUN is appropriate for the host, since it combines a powerful development environment with a suitable graphics capability to display results.

We have been using a SUN 3/280 server and SUN 3/60 workstation as the host machine for our prototype machine The SUN 3/280 contains a Transputer based interface card which communicates via a fibre optic cable to the symbolic Transputer layer.

4.2 Symbolic Processing Layer

The symbolic layer provides the top level of processing, dealing with abstract model data, and is an MIMD array of coarse grained processors. The processor we use for this is the INMOS Transputer, which is specifically designed with MIMD multiprocessing in mind. It includes hardware support for inter-processor communications in the form of four DMA driven serial links which can communicate at up to 20Mbits per seconds each without processor intervention. It also provides micro-coded support for multi-tasking which allows very fast context switching, and transparently handles interaction with the on board DMA controlled links.

The Transputer supports a model of concurrency known as Communicating Sequential Processes or CSP defined by Hoare [7]. This models breaks problems up into a set of independent sequential processes which communicate and synchronise only by means of channels. The channel is a unidirectional data stream which causes a rendezvous for every data item passed through it. Thus if a process writes to channel it is suspended until another process reads the channel. Conversely if a process reads from a channel it will be suspended until data has been written into the channel by another process.

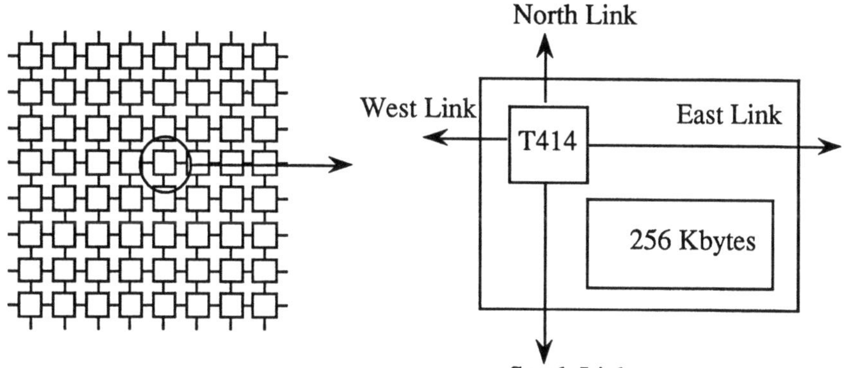

Figure 3: Symbolic Processor Array

The Transputer has instructions which directly implement these channel semantics, regardless of whether the channel is between two processes on a single Transputer or over a link between two processes on two different Transputers. Using this mechanism it is possible to write programs that are independent of the number of Transputers in the system provided that there are at least as many processes in the system as there are processors.

The Transputer is available both with and without hardware floating point support. Ideally we would like to have used the T800 floating point Transputer, but because of financial constraints our prototype hardware uses T414 non-floating point device. Our symbolic processing layer consists of an 8 x 8 array of N-S-E-W connected T414 Transputers each with 256Kbytes of RAM. We also have a Transputer based frame-buffer used for displaying processed images. This layer has been designed and constructed by us, and has proved a useful vehicle for software development, as well as providing the first stage in our prototype pyramid machine.

Each Transputer will communicate with the Cluster Controller below it by means of shared memory. In our prototype hardware this is 2Kbytes of dual port static RAM which can be accessed both by the Transputer and the cluster controller. This memory is used by the symbolic processor to send commands to the cluster controller, as well as being used for the bidirectional transmission of data between the symbolic processor and cluster controller.

4.3 Controller Array

The controller array is an 8 x 8 array of cluster controllers . Each cluster controller is responsible for generating the instruction stream for one cluster, which consists of the cluster controller and a 16 x 16 region of the iconic layer. Each cluster is equivalent to a conventional SIMD machine, and it is this that gives the machine its Multi-SIMD capabilities. The cluster controller is also responsible communicating data to and from the iconic layer.

The controller consists of a micro-instruction sequencer which fetches one micro-instruction at a time and broadcasts it to all of the processing elements (PE's) within the cluster. Each instruction specifies the operation that the PE's ALU is to perform, and also the source and destination of its operands. These can be either internal registers or locations in iconic layer memory. Since all the PE's operate on the same location in memory at the same time it can be thought of as acting on a plane of data.

The operation for the PE ALU is specified in the cluster controller micro-instruction, but the source and destination for the operands are generated dynamically which allows the same PE ALU operation to be performed on several different locations in memory. This is particularly useful for implementing multi-bit operations.To allow this the cluster controller includes a scaler ALU for address calculations. This ALU is also used to generate the conditions needed to control the flow of the micro-instruction sequencer.

To achieve maximum performance from the iconic layer it is necessary to provide one micro-instruction on every clock cycle. This proves very hard to achieve since for each micro-instruction several scaler operations may be required to calculate the operand addresses. However we have managed to approach this level of performance by allowing the address calculations to proceed in parallel with the execution of iconic layer PE instructions.

The cluster controller is also responsible for communication of data between the symbolic layer and the iconic layer. Information is passed between the controller and the symbolic processors using the dual ported memory, and between the iconic layer and the controller via a set of hardware registers that will be described later.

A functional diagram of the cluster controller element is shown in Fig. 4. It consists of a micro-programmable 16 bit processor consisting of an AMD 29116 ALU and an AMD 29331 Sequencer, with 16K x 68 bit micro-code store and 2K 16 bit dual ported data memory. These are connected together via a single 16 bit data path. The dual ported memory is shared between the cluster controller and the symbolic processor and is used for communication between the two. At boot time the host processor loads the cluster controller micro-code store with a standard set of micro-routines, which can later be called up through a software protocol using the dual ported memory.

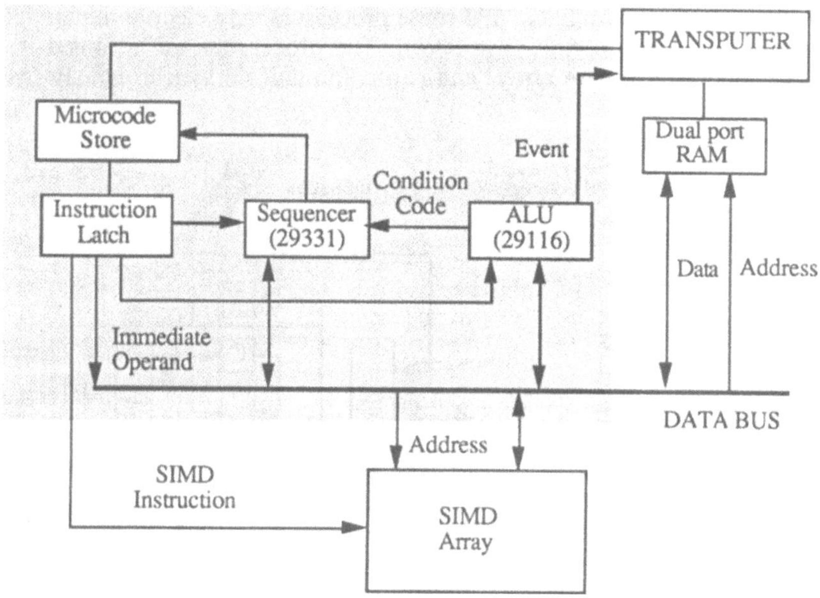

Figure 4: Cluster Controller

The cluster controller's micro-code word is divided into four fields one each for the ALU, sequencer, data path and the iconic processor array. Thus on each cycle the cluster controller can control all four elements in parallel. In this way it is possible to overlap address calculations performed by the 16 bit ALU and pixel instructions performed by the iconic layer, and so maintain the maximum possible iconic level instruction throughput.

The cluster controllers are capable of Multi-SIMD operation with each cluster operating independently from each other. However it is also possible to operate several clusters as a single entity if the iconic processors in adjacent clusters must be synchronised in order that communication across the cluster boundaries performs as expected. This synchronisation is performed by the cluster controller level, using inter-cluster controller communication links to rendezvous synchronised instructions.

4.4 Iconic Layer

The iconic layer performs the low level pixel processing, such as intensity interpolation, hidden surface elimination, convolutions etc. and consists of an SIMD array of bit-serial processors. The total array size is 128x128 processors, but this is broken up into an 8x8 array of 16x16 regions, each one controlled by its own controller. Each 16x16 region and its controller form one cluster. All the processors in one cluster execute the same instruction stream, the only exception being that some processors may elect to remain idle instead of executing a certain instruction. The processors are arranged in a N-S-E-W connected square array and can communicate synchronously over this network.

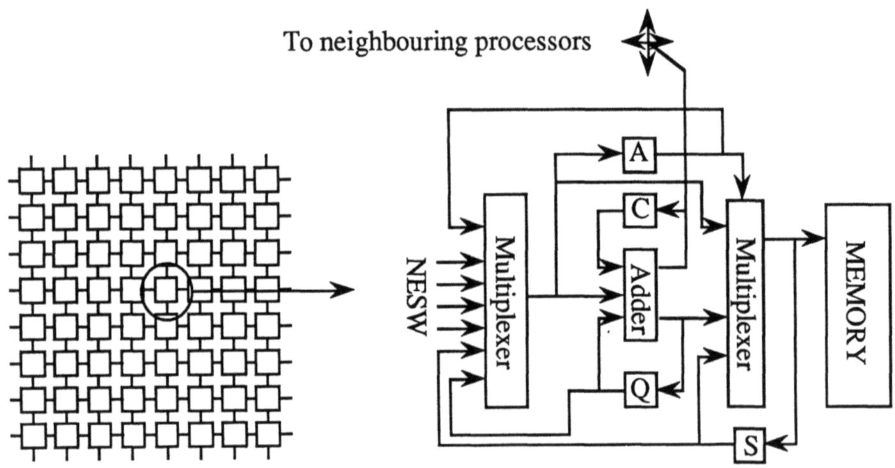

Figure 5: DAP Based SIMD Array

The processors used for this layer are the AMT Distributed Array Processors. The AMT DAP chip integrates 64 bit serial processors onto a single chip. Each processing element consists of a one bit ALU, four one bit registers, and some private external memory. The ALU provides a variety of one bit arithmetic and logical operations with add with carry being the most complex. Since the instruction stream is provided externally there are no control instructions in the DAP instruction set. Most instructions are arithmetic or logic operations, with a small number of communication instructions. Each processor can perform operations on registers and operations which require only a single memory reference in a single clock cycle. Operations which require two memory references take two cycles.

In addition to the processing array each chip incorporates an edge register, which can be used to access a whole row or column of processors in one go. This edge register is mapped into the register space of the cluster controller, and can be used under software control to transfer data to and from the DAP processing elements. Our design also features a mechanism for counting the number of processing elements that respond to a certain key, to allow us to implement associative algorithms efficiently.

In our prototype machine each 16 x 16 cluster is implemented using four DAP chips. Each processing element has 32Kbits of fast static memory making a total of 1Mbyte of memory per cluster.

4.5 Multiple-SIMD Design

A particularly important aspect of our architecture is its Multi-SIMD organisation, where the SIMD array is divided into independently controllable clusters. Other machines have been suggested that combine MIMD and SIMD parallelism [9] but they suffer from two major flaws which the Multi-SIMD approach overcomes.

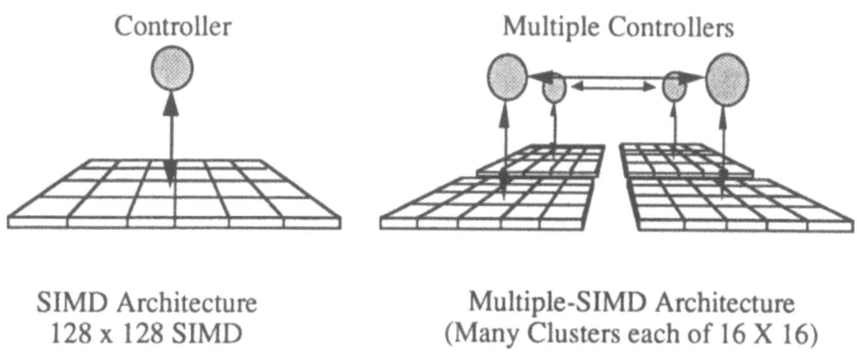

Figure 6: SIMD vs. Multi-SIMD

For a combined MIMD/SIMD machine to work effectively it must be able to pass data between its two halves quickly, which implies a high bandwidth connection. In a conventional SIMD machine the single controller becomes a bottleneck for communication and limits the bandwidth. In our design multiple controllers allow a far greater vertical bandwidth than would otherwise be possible. This approach also scales more readily since the number of controllers increases as the size of the array increases.

The other main advantage of a Multi-SIMD design is increased local autonomy. In a conventional SIMD machine if the processing is concentrated in a small area of the image the rest of the array has to lie idle. If the 128 x 128 array of processors were rendering a polygon, which probably only covers in the order of 100 pixels then only 1 % of the array would be active, or put another way the array would be operating at only 1% efficiency. With a Multi-SIMD design it is possible for each cluster to work on a different area of the image independently of all the others. Thus, taking the previous example, since the area of one cluster is of a similar order to the polygon, we might expect each cluster to be able to work on a different polygon, giving a factor of 64 (the number of clusters) improvement in performance. In practice of course this will not be fully realised but nonetheless a significant improvement is to be expected.

4.6 Current Status

We have built and tested a prototype machine. This consists of a full 8 x 8 array of T414 Transputers each with 256Kbytes of memory on multi-layer printed circuit boards. Connected to this is a single cluster. The cluster consists of a cluster controller which is a single wire wrapped board, and a 16 x 16 array of DAP processing elements implemented as four DAP chips with 1 Mbyte of fast external memory, implemented as two multi-layer printed circuit boards. The Transputer array runs at 20MHz while the cluster runs at 10MHz.

5. Programming

The pyramid architecture consists of three different processor types (Transputers, bit-slice, DAP PE's), each of would normally be programmed its own language. Ultimately it is hoped that a user will see only one language, namely Parallel C++, but at the moment two main languages are used, Parallel C and cluster assembly language.We will now describe these languages and then go on to discuss our future plans to integrate them.

5.1 Transputer Parallel C

This is a commercial product and so we will touch on it only briefly. It provides an ANSI C compiler and a set of function calls to implement the OCCAM style of concurrency, namely communicating sequential processes or CSP. This includes the ability to create processes and to communicate over channels. Processes and channels can be created dynamically and the language allows recursion.

Parallel C is used to program the symbolic processors. Typically the programs on the symbolic processor will coordinate the operation of their associated cluster. We have written a library which provides an interface to a set of routines written in cluster assembly language which run on the cluster controller. These routines perform a variety of functions from low level operations such as multi-bit arithmetic to higher level functions such as scan converting of a polygon. The C functions invoke a remote-procedure call mechanism which makes use the dual-ported memory to write into the cluster controller's instruction queue. The cluster controller takes instructions from this queue and executes the appropriate micro-code routine. Once it has completed this is writes the results into dual-ported memory and generates an event on the symbolic processor. The Transputer's scheduler receives the event and can then restart the process. This mechanism allows the symbolic processor to do other computations while the cluster controller is busy.

5.2 Cluster Assembly Language

The cluster assembly language provides a low level interface to the devices which make up the cluster, namely the scaler ALU, the micro-instruction sequencer, the data path and the DAP array . The cluster assembly language is effectively two assembly languages in one, namely AMD 29116/29336 bit-slice assembler and DAP assembler. Where possible the same mnemonics have been used in the assembler as used by the manufactures of the devices. This does lead to a somewhat irregular syntax, but it means that the manufacturers documentation still makes sense in the context of our machine.

The cluster controller design allows each micro-instruction to fully specify the operations of all four functional units, and the cluster assembly language reflects this by using four fields in each instruction, one for each functional unit, so that each line of assembly language corresponds to one 68-bit micro-instruction.

The use of a wide instruction word, with its inherent parallelism provides a worthwhile performance increase, but it does make programming the machine quite difficult. It is up to the programmer to work out which instructions can safely be carried out in parallel, and since they are all highly interdependent this is no easy task. We intend to automate this process in the future by providing a high level language compiler for the cluster controller but for the time being it is strictly a real programmers machine !

5.3 Parallel C++

We would ideally like to completely integrate the programming of our machine by providing a single language which can express both SIMD and MIMD models of concurrency. It is our hope that C++ will allow us to achieve this. Object oriented programming, which breaks programs into self contained units of data and program, has already been used effectively as a parallel programming technique for MIMD machines [10], and it is our feeling that it could be used equally well to implement the model of SIMD parallelism used in DAP FORTRAN. We intend to use the object oriented facilities in C++ to generate a suitable class library which will include both MIMD and SIMD objects, and thus provide the integrated programming environment we need.

6. Polygonal Mesh Rendering

Having described the architecture and its programming we will now look at mapping a specific graphics application onto it, namely the standard polygon mesh rendering pipeline.

A polygonal mesh rendering systems takes a model, defined by planar polygonal patches, and processes it in stages to produce an image of the model. The implementation has two main objectives. The first is to map each stage in the process onto the processor which can most efficiently implement it. To do this we need to look at the type of calculations that each stage performs, and also the structure of the communication at each stage. In addition to this we would like to use algorithms at each stage that are arbitrarily scalable. This means that as the number of processors increase so does the throughput of each stage.

The rendering is carried out in a number of separate stages, which can be arranged in a pipeline, and indeed this is generally how dedicated hardware implementations are arranged [12]. The early stages of the pipeline, which consist of database control, 3D transformations, clipping and shading calculations, are mapped onto the symbolic processor, since they involve complex operations, usually floating point, performed on relatively few data items. The later stages which consist of scan conversion, hidden surface elimination and intensity interpolation, are mapped onto the low level clusters, since they involve much simpler integer calculation which can be applied to all the pixels in each polygon simultaneously.

Previous attempts at mapping the rendering pipeline onto arrays of MIMD processors have involved using one processor for each stage in the pipe, emulating the approach used by custom hardware [8]. This approach has three disadvantages, first it does not scale readily, which is one of our primary aims, since the number of pipeline stages is fixed to the number of stages in the rendering process. Secondly this approach involves the polygon data being transmitted between each stage in the pipeline, which can be done in a hardware pipeline with no overhead, but in an MIMD machine where communications are a significant overhead this is a problem. Thirdly the pattern of communication implies that the processors be arranged as a pipeline, which does not correspond to our architecture.

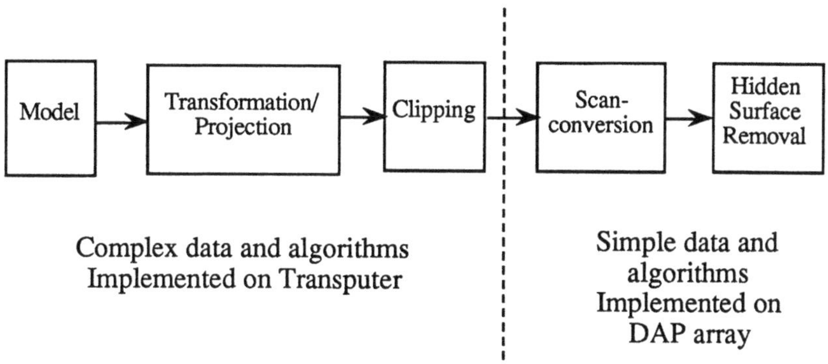

Figure 7: Polygon Rendereing Pipeline

The approach we have taken is to use a static 'processor farm'. This involves distributing the polygon data base evenly amongst the available Transputers, such that each processor has an equal number of polygons. The viewing parameters and lighting details are then broadcast to the processors, which calculate all the transformations and shading for their set of polygons. The polygons are then clipped to cluster boundaries, which may involve polygons which span more than one cluster being broken into several pieces. These clipped polygon are then sent to the symbolic processor associated with the clusters which contains them. This processor then passes them onto the cluster controller for rendering. We have found that this technique provides nearly linear speed up with number of processors, and can be extended indefinitely, up to one processor per polygon. It is also topology independent.

This method relies on being able to send polygon data from one processor to any other in the array, so that clipped polygons can be sent to the appropriate symbolic processor for rendering. To allow this we have implemented a message based router, which can route addressed packets from one processor to any other. The existence of this router has other advantages, in that it allows any node to print data on the host machine, which is extremely useful for debugging purposes.

Once the clipped polygons arrive at the appropriate cluster controller they must then be scan converted. Clearly we wish to render all the pixels in each polygon in parallel, and so the conventional incremental algorithms are not suitable. Instead we use an algorithm developed by Fuchs [11] for his Pixel Planes machine.

The algorithm proceeds in a number of stages, taking one edge of the polygon to be rendered at a time. For each edge the pixel processors in the patch calculate the equation Ax + By + C, where x and y are the coordinates of the processor within the patch, and A, B and C are the coefficients of the line Ax + By + C = 0 which corresponds to the edge. The coefficients are calculated such that pixels outside the polygon return a consistent positive or negative result, so that all the pixels which do not fall inside the polygon can be eliminated. If this process is repeated for all the edges, those pixel which have not been eliminated must be inside the polygon.

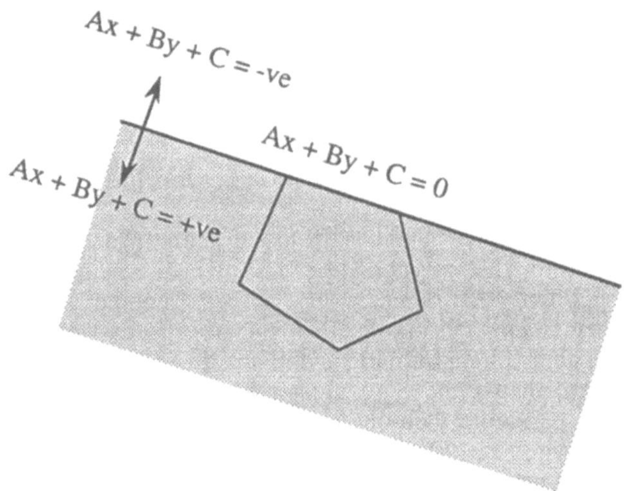

Figure 8: Interior Point Determination

Once the pixels within the polygon have been determined, the next step is to determine the distance of each pixel from the viewer, so that hidden surface elimination can be performed. To do this the processors evaluate the same Ax + By + C equation, but this time the coefficients are of the plane Ax + By + C = z, where z is the depth of the polygon from the view at the point x,y. Each processor then compares this value with any previously stored values for polygons already plotted in that pixel, and determines whether the new value is closer to the viewer than previous values. Finally all those pixels which are inside the polygon and closer to the viewer than previous polygons are set to the colour value for that polygon. This value may itself be calculated using a linear equation to achieve Gourard shading.

6.1 Performance

The Transputer array is expected to be able to process 400,000 triangles per second, assuming that T800 Transputers are used. A single cluster can render approximately 20,000 triangles per second. A full sized SIMD array could process up to 1.2 million triangles per second, but this assumes that all clusters are fully active the whole time which is optimistic. However we would hope that it would able to render at the a rate similar to that produced by the Transputer Array, giving a total system performance of 400,000 triangles per second.

7. Conclusions

We believe that a programmable parallel machine can provide the level of performance normally only available with custom hardware while not restricting the user to one specific set of algorithms. However this relies on achieving a good match between hardware and software, which can only be achieved by targeting the machine at particular problem area where all the algorithms share a similar problem structure. We have designed a machine which is oriented towards image based problems, such as image understanding and computer graphics. These problems are generally difficult to map onto existing parallel architectures because they contain many data structures which map most effectively on to widely different machines. Our architecture overcomes this by incorporating both coarse grained MIMD and fine grained SIMD processors.

Acknowledgements

This work is supported in part by contract N00014-87-G-0241 from the U.S. Innovative Science & Technology Office (addministered by Dr K.Bromley, Office of Naval Research).

References

1. K.E.Batcher, "Design of the Massively Parallel Processor", *IEEE Trans.* **C-29**, pp 836-840,1980

2. M.J.B.Duff, T.J.Fountain (Editors), "Cellular Logic Image Processing", Academic Press, New York, 1986

3. D.J.Hunt, "AMT DAP - a processor array in a workstation environment", *Computer Systems Science and Engineering,* 4 (2), pp107-114, April 1989

4. T.J.Fountain, "A Review of SIMD Architectures", in *Image Processing System Architectures*, ed J.Kitler.

5. M.J.Flynn, "Very High Speed Computing Systems", *Proc IEEE,* **V54** No. 12, 1966.

6. G.R.Nudd et al," WPM: A Multiple -SIMD Architecture for Image Processing", *IEE Conference on Image Processing and Applications*, Warwick, July 1989.

7. C.A.R.Hoare, "Communicating Sequential Processes", *Prentice Hall*, 1985.

8. T.A.Theoharis, "Exploiting parallelism in the Graphics Pipeline", *Technical Monograph PRG-54*, University of Oxford.

9. I.Page, "DisArray: A 16x16 RasterOp Processor", *Eurographics '83.*

10. W.Bronnenberg, "POOL and DOOM", *PARLE '89,* Lectures Notes in Computer Science, Springer-Verlag

11. H.Fuchs, "PIXEL-PLANES: A VLSI-oriented design for a raster graphics engine", *VLSI Design* **VII**, no.3.

12. J.Clark, "The Geometry Engine: a VLSI geometry system for graphics". *Computer Graphics (SIGGRAPH '82 Proceedings)*, **vol 16**, no3, pp127-133.

13. E.F.Gehringer, "A Survey of Commercial Parallel Processors", *Computer Architecture News,* **Vol16**,. no4, 1988.

14. W.D.Hillis, "The Connection Machine", MIT Press, 1985.

Two-level Pipelining of Systolic Array Graphics Engines*

J. A. K. S. Jayasinghe and O. E. Herrmann

Twente University
Laboratory for Network Theory
P.O. Box 217
7500 AE Enschede
The Netherlands

In a systolic array, the maximum operating speed is determined by the most complex operation performed. In a systolic array graphics engine, capable of generating high quality images, one has to perform complex operations at a very high speed. We propose to use pipelined functional units in systolic array graphics engines as they can perform complex operations at high speeds. Due to time-varying discontinuities of operations performed by systolic array graphics engines, introduction of pipelined functional units is a complex problem. In this paper we present a methodology which solves this problem by a graph-theoretic approach. Furthermore, we characterize the architectures which can be improved by pipelined functional units.

Categories and Subject Descriptors:
B.7.1 [Integrated Circuits]: Types and Design Styles - VLSI
C.1.1 [Single Data Stream Architectures]: Pipeline Processors
C.3 [Special-purpose and Application Based Systems]: Real-time Systems
I.3.1 [Computer Graphics]: Hardware Architecture - Raster Display Devices
I.3.7 [Computer Graphics]: Three-dimensional Graphics and Realism - Color, Shading, Shadowing and Texture

Key Words & Phrases: Raster Graphics, Computer Graphics, Real-time Scan-conversion, Systolic Arrays, Two-level Pipelining.

*The work described in this paper is a part of a project for development of a graphics workstation. The project is sponsored by the Dutch Research Foundation (STW), under contract CWI77.1249, carried out by Twente University, Enschede, The Netherlands and Center for Mathematics and Computer Science, Amsterdam, The Netherlands.

1 Introduction

Raster graphics systems have become very popular over the past decade, due to the image quality achieved by rendering the complete object instead of rendering the outline of the object as in vector graphics. Though raster graphics systems are capable of rendering high quality pictures, they need several orders of magnitudes higher processing power for image generation compared to vector graphics systems. As the display must be refreshed continually, almost all raster graphics systems decouple the image generation and display refresh process by an intermediate buffer called *frame buffer*. Though a typical frame buffer needs several Mbytes of memory, cheap memories made the frame buffer concept feasible. The frame buffer is organized as a pixel array for fast screen refresh. Although, the structure of the frame buffer is extremely suitable for screen refresh, its structure is not well suited for fast interaction [12]. In [12], it has been argued that the frame buffer must be replaced by a structured object list tuned for both fast interaction and screen refresh. In this case, pixel values have to be calculated in real-time. Some graphics systems can calculate the pixels from an object description [1],[2],[3],[5],[6], which could be adapted to operate with structured object lists. Systems like Pixel-planes [5], add pixel processing logic to each pixel memory location and systems like PROOF [6] use one processor-per-object. Addition of pixel processing logic to each pixel results in a multi-fold increase in overall size of the system, since the size of the processing logic is several times the area of a memory cell. On the other hand, practical object processors can handle only very simple objects like triangles. Therefore, one needs a very large number of object processors to get a complex image. As these approaches produce very big systems, one can question the feasibility of replacing the conventional frame buffer by a structured object list.

Fortunately, the overall system size of the multi-processor systems reported in [1]-[3], is much smaller than systems based on processor-per-pixel or processor-per-object approaches. These multi-processor systems are based on a one-dimensional systolic array. They can generate the images at the display refresh speed without a frame buffer. In this paper we refer to these graphics engines as SAG (*Systolic Array Graphics*) engines. In SAG engines, pixel storage has been limited to a row of pixels, which has been distributed over the processor array. In general, a systolic array consists of an array of identical processor elements (PEs), and connections between PEs have been restricted only to near neighbors. Hence, systolic arrays are very attractive for VLSI implementation and furthermore, communication between PEs is fast and cheap.

When all the PEs are finished with their computations, a global clock enables the communication between PEs. Therefore, throughput of the systolic array is determined by the delay of the most complex operation supported. Hence, we can expect a reduction in maximum operating speed when complex operations are added to an SAG engine to provide better image quality. The maximum operating speed of the SAG engines reported in [2],[3], have been limited to 40-50ns due to (16 bit fixed-point) addition, even though advance IC technologies like $1.2\mu m$ CMOS have been used. A 50ns pixel clock cycle is enough for a 512x512 pixels display refreshed at a rate of 50Hz. A 1024x1024 pixels display refreshed at the same rate needs a 12ns pixel clock cycle. Computer graphics users have ever increasing desire for higher resolution displays, better image quality, and higher interaction speed. Therefore, SAG engines will not be feasible for tomorrow's computer graphics, unless one has some means to speed up the operations performed in these graphics engine.

In the past, speed of the ICs have been increased dramatically by scaling the device sizes. Early integrated circuits (circa 1965) had minimum feature sizes in the range of 10 to 20μm. Today, sub-micron minimum feature sizes are used in the state-of-the-art IC technology. When the parasitic wiring capacitance is neglected, scaling of minimum feature sizes by factor S reduces the delay by factor S^2 (theoretically) [7]. But the feature sizes of current ICs are such that speed improvements due to scaling are lower than the above theoretical value due to wiring capacitances and other parasitic effects. The parasitic effects become more and more dominant when the minimum feature sizes become smaller and smaller. However, there is a fundamental physical limit on the transistor sizes, and according to [7] this is of 0.25 μm order. Based on the above facts (speed of scaled IC technologies, speed requirements of higher resolution displays and speed of reported SAG engines) one can conclude that bare speed-up of scaled IC technology is not enough for tomorrow's computer graphics hardware based on SAG engines.

Pipelined functional units have been used in computer hardware to achieve higher throughputs. A systolic array built from pipelined functional units has been first reported in 1982 [11], which has been referred to as *two-level pipelined systolic array*. Hence, in this paper we propose to use pipelined functional units in SAG engines to increase the operating speeds. The introduction of pipelined functional units into an SAG engine is a complex problem, as functions performed by PEs in an SAG engine are subject to time-varying discontinuities. (This is explained in detail in the last paragraph in Section 2.) But, as we can see in this paper, the complexity can be handled by a systematic approach.

The primary goal of this paper is to make a framework to derive two-level pipelined SAG engines from a non (two-level) pipelined SAG engine. In the next section, we present a survey of working principles of SAG engines reported in literature. In Section 3, a formal approach is presented, which can be used to derive two-level pipelined SAG engines from a non (two-level) pipelined SAG engine. In Section 4, some issues on pipelinability are discussed. An architecture of a two-level pipelined SAG engine is presented in Section 5. Finally some conclusions are drawn in Section 6.

2 Principles of SAG Engines

The SAG engines are built from a one-dimensional identical processor array as shown in Figure 1. According to the directions of data movements, we can categorize the SAG engines into two groups which will be referred to as *contra-flow* and *contra-flow free* architectures. In contra-flow architectures, video information travels in the opposite direction to data and instruction flow. Video, data and instructions travel at the speed of one processor per cycle. The difference in contra-flow free architectures is that video information is sent to the display through a global bus. The architectures presented in [1],[2] are contra-flow architectures whereas the architectures presented in [3],[4] are contra-flow free architectures.

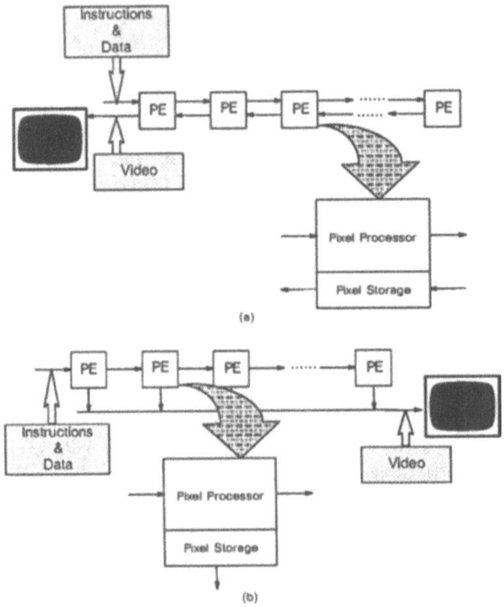

Figure 1: Two Different SAG Engins: (a) Contra-flow, (b) Contra-flow free.

In contra-flow architectures only half of the processors contain the instructions at any cycle. If processors at even locations contain the instructions on cycle N, then processors at odd locations idle on cycle N because only the processors containing instructions are active. As instructions travel one processor per cycle, the states are reversed on the next cycle. However, in contra-flow free architectures, all the processors are active at any cycle. Apart from these differences, contra-flow and contra-flow free architectures have the following common behavior: each processor array is fed by instructions, which contain the information of a span of pixels, along with some processor addresses indicating some important points in the span. Different functions are performed by sending different instructions. For example, constant shading or Gouraud shading is done by sending different instructions. When the instructions pass through each processor, the pixel values are calculated by the processors when they are addressed by the address field and stored in a local pixel storage. Furthermore, when the processor is addressed by the instruction, some data associated with each instruction are also modified before the instruction is sent to the next processor.

Due to real-time requirements, the processors in the array work in two modes: *refreshing and non-refreshing*, which is determined by the instruction on each processor. At any time only one processor can be in refreshing mode. When a processor is in the refreshing mode, the content of the pixel storage is sent to the display. Non-refreshing processors do the manipulations on the data stream, if they are addressed by the address field of the incoming data. Furthermore, the non-refreshing processors on the right of the refreshing processor operate on the current pixel row, while the non-refreshing processors on the left of the refreshing processor operate on the next pixel row. As instructions travel one processor per cycle, in each cycle the processor in refreshing mode switches back to non-refreshing mode while its right neighbor becomes the next refreshing processor. Therefore, SAG engines process the pixel information in a pipelined fashion without any exclusive pipeline filling or draining. Table 1 shows the instructions supported by different SAG engines reported in literature.

The functions performed by PEs in an SAG engine are subject to time-varying discontinuities. For example when the Z-buffer algorithm is executed on an SAG engine, the Z value of the polygon closest to the eye so far encountered is stored in a register for a variable number of clock cycles. The number of clock cycles depends on conditions such as whether the next instruction has a pixel span which overlaps with the pixel location assigned for the processor, or whether the new pixel span is closer than the previous spans. As the instructions for the Z-buffer algorithm pass through each PE, the Z value associated with each instruction is updated only by those processors that are assigned to a given pixel span, which introduces another form of functional discontinuity. Note that in a systolic array designed for convolution, matrix multiplication, etc., no functional discontinuities can be seen as all the PEs perform the same function. When pipelined functional units are introduced, one must make sure that no empty cycles, no exclusive pipeline filling or draining cycles are introduced to handle the functional discontinuities, because they reduce the number of advantages we have in an SAG engine.

3 Designing Two-level Pipelined SAG Engines

3.1 Related Work and Their Strengths and Weaknesses for Two-level Pipelining of SAG Engines

Designing a two-level pipelined systolic array can be done in two ways: transforming a non (two-level) pipelined design systematically into a two-level pipelined design, or designing the two-level pipelined systolic array from scratch. The later approach has been demonstrated in literature [10] in applications such as computing inner products. To our opinion this approach is not well suited for SAG engines due to the complexity introduced by time-varying functional discontinuities. The former approach was first introduced by Kung et al. [8]. In fact, in two-level pipelining, one tries to improve the clock period by introducing pipeline registers into critical paths. A technique called *re-timing* which could optimize the clock speed of a synchronous circuit was introduced by Leiserson et al.[9]. In re-timing, the clock speed is optimized by re-locating the registers in a given circuit while the logical behavior of the system is kept intact. Graph-theoretic approaches have been used for re-timing and two-level pipelining by Leiserson et al. and Kung et al. In these graph-theoretic approaches, combinational logic and communication links are denoted by the nodes and edges of a graph. The number of registers on a communication link is denoted by a weight on the corresponding edge. The weights on nodes represent the propagation delay of the corresponding combinational logic, which has been omitted by Kung et al. In both approaches, registers introduce fixed latencies. In an SAG engine, the latencies introduced by registers are subject to time-varying discontinuities. Therefore, both approaches cannot be used to improve the clock speed of an SAG engine.

In Reference	Instruction	Description
1	REFRESH( )	Send the local pixel storage to the display and reset the processor.
	EVAL0(X,DX,I)	Store value I in processors X,X+1,...,X+DX . (Constant shading)
2,3	REFRESH( )	Send the local pixel storage to the display and reset the processor.
	EVAL0(X,DX,I,Z,DX)	Calculate the intensity and depth by zero and first order interpolation at processors X,X+1,..X+DX and store the calculated intensity and depth if stored depth is larger than calculated depth. (Constant shading and hidden surface removal by the Z-buffer algorithm, supported only in 3.)
	EVAL1(X,DX,I,DI,Z,DX)	Calculate the intensity and depth by first order interpolation at processors X,X+1,..X+DX and store the calculated intensity and depth if stored depth is larger than calculated depth. (Gouraud shading and hidden surface removal by the Z-buffer algorithm.)
4	REFRESH( )	Send the local pixel storage to the display and reset the processor.
	EVAL0(X,DX,I)	Accumulate value I in processors X,X+1,...,X+DX until next refresh. (Constant shading)
	EVAL1(X,DX,I,DI)	Calculate the intensity by first order interpolation at processors X,X+1,...,X+DX and accumulate the calculated intensity until next refresh. (Gouraud shading)
	EVAL2(X,DX,I,DI,DDI)	Calculate the intensity by second order interpolation and accumulate the calculated intensity until next refresh. (for Phong Shading)
	SETI(X,DX,I)	Change the intensity at processor locations X,X+DX,X+2DX,... to I during next interpolation.
	SETDI(X,DX,DI)	Change the first derivative of the intensity at processor locations X,X+DX,X+2DX,..., to DI during next interpolation.
	SETDDI(X,DX,DDI)	Change the second derivative of the intensity at processor locations X,X+DX,X+2DX,... to DDI during next interpolation.
	DIS(X,DX)	Disable the accumulation between processor locations X and DX.
	NOP( )	Do nothing

Table 1: Instructions Supported in Different SAG Engines.

3.2 A New Methodology

In this section, we present a new graph-theoretic approach which can be used to convert a systolic array (built from non-pipelined functional units) with time-varying functional discontinuities into a two-level pipelined version. In fact this approach can be used not only for SAG engines but also for any systolic array with or without functional discontinuities.

In our methodology, the original systolic array is represented at bit-level by a finite, vertex-weighted, edge-weighted, directed graph $G = \langle V, V', E, d_V, d_{V'}, w\rangle$, (from now on, for simplicity we say *graph.*) where V, V' and E denote the nodes and edges of the graph. The labels $d_V, d_{V'}$ and w denote the weights on the nodes and edges respectively. The graph G which is constructed by replicating the graph of a PE, $G_{PE} = \langle V_{PE}, V'_{PE}, E_{PE}, d_{V_{PE}}, d_{V'_{PE}}, w_{PE}\rangle$ and connecting vertices according to communication between PEs as the entire array is built from identical PEs.

In order to construct the graph G_{PE}, the bit-level functional units are represented by vertices V_{PE} and bit-level storage by vertices V'_{PE}. The communication between nodes are denoted by the edges and for each edge $e \in E_{PE}$, $w_{PE}(e)$ denotes the earliest communication time slot for all legal combinations of instructions. The edges which communicate in the $i^{th}(i = 0, 1, 2, ...)$ time slot are weighted by i. Each vertex $v \in V_{PE}$ is weighted by the numerical propagation delay of the functional unit $d(v)$. Each vertex $v' \in V'_{PE}$ is multiply weighted by $d(v'_{a \to b})$ for each input edge a and output edge b connected to node v', when there is a direct information flow[1]. The quantity $d(v'_{a \to b})$ indicates the minimum latency (which is under control of instructions) between information flow from edge a to b through the storage node. If there is no direct information flow from edge a to b then $d(v'_{a \to b})$ is undefined.

The maximum clock speed of the circuit is determined by the propagation delays of the critical paths. A critical path in our graph G can be identified as a directed path activated in the same time slot such that the sum of the node weights on that path is maximum over the entire graph. (If a critical path goes through a storage node, the critical path is terminated at the storage node whenever the vertex weight corresponding to that path is non-zero, as the vertex weights of storage nodes denote the latency whereas the weights of other nodes denote the propagation delay.) In order to improve the clock speed the graph is re-timed. In re-timing, pipeline registers are added to critical paths to meet the given speed requirements, and then some more additional pipeline registers are added to some other edges and/or latencies at some storage vertices are changed such that the following theorem is satisfied.

Two-level Pipelining Theorem: If a two-level pipelined design is obtained by adding pipeline registers to some edges and/or changing the latencies of storage nodes, the logical behavior of the system will be kept intact if the differences in latencies through all pairs of paths between any two nodes are equal in the original and re-timed graphs.

[1]In Figure 5 the information flow from edge a_i to a_{i+1} is direct whereas the information flow from edge a_i to d_i is indirect. There is no information flow from edge a_i to b_{i+1}

Proof: Let m, n be two nodes and p_1, p_2 be two paths from m to n. Let path p_1 be activated on $i_{p1_1}, i_{p1_2}, ..., i_{p1_{N1}}$ time slots ($i_{p1_1} < i_{p1_2} ... < i_{p1_{N1}}$) and p_2 be activated on $i_{p2_1}, i_{p2_2}, ..., i_{p2_{N2}}$ time slots ($i_{p2_1} < i_{p2_2} ... < i_{p2_{N2}}$). If we introduce a latency of k cycles in path p_1, by introducing pipeline stages and/or changing storage latencies, node m gets the data from node n through path p_1 in time slot $i_{p1_{N1}} + k$. In the original graph node m gets the data from node n through paths p_1 and p_2 in time slots $i_{p1_{N1}}$ and $i_{p2_{N2}}$ respectively. In order to keep the logical behavior of the graph intact we must get data through path p_2 in time slot $i_{p2_{N2}} + k$. Now, we can see the difference in latency through path p_1 and p_2 is equal to $(i_{p1_{N1}} - i_{p2_{N2}})$ cycles in both original and re-timed graphs. This argument can be applied to all paths between any node pair to prove the theorem. □

It can be proven that the application of the two-level pipelining theorem produces $|E| - |V| + 1$ linearly independent equations, where $|E|, |V|$ are the number of edges and vertices in the graph [13]. These equations contain $|E| + |V'_{\geq 1}|$ unknowns where $|V'_{\geq 1}|$ is the number of non-zero weights in the storage nodes. As we have more unknowns than equations, the re-timing problem can be solved by linear programming, for example minimizing the total register count as Leiserson proposed [9].

3.2.1 An Example

Figure 2(a) shows a systolic array finite impulse response (FIR) filter[2] whose output is given by $y_i = \sum_{k=0}^{k=1} x_{i-k}$. In this filter input x_i and output y_i travel to the left and right respectively at a speed of one processor per cycle. As y_i travels, it collects the partial sums and hence the complete sum is formed at the rightmost PE. For the proper operation, the input x_i must supplied at every other cycle. Hence, output y_i is also produced at every other cycle. Figure 2(b) shows the graph of the systolic FIR filter when a carry ripple adder is used for addition. For simplicity, we have assumed that calculations are performed by 2 bit binary numbers. The nodes v'_1, v'_2 (v'_3, v'_4) represent the delay D which stores x_i in PE1 (PE2). Similarly, nodes v'_{11}, v'_{12} (v'_7, v'_8) represent the delay D which stores y_i in PE1 (PE2). The adder in PE1 (PE2) is represented by nodes v_9, v_{10} (v_5, v_6).

Let us assume that $k_{i,j}$ denotes the number of pipeline registers placed on edge (v_i^*, v_j^*) [3] and k_i denotes the non-zero latency at storage node v'_i for a two-level pipelined design. Let us place a single pipeline register in each edge $(v_9, v_{10}), (v_5, v_6)$ as these edges belong to the critical paths, i.e. $k_{9,10} = k_{5,6} = 1$. Note that these edges represent the carry propagation in the adders. Application of the two-level pipelining theorem generates 5 linearly independent equations as follows.

---

[2] In fact this systolic array has no functional discontinuities. But we use this example for simplicity.

[3] Here v_i^* stands for v_i or v'_j, as nodes in the example are numbered such that v_i and v'_j exists only for $i \neq j$.

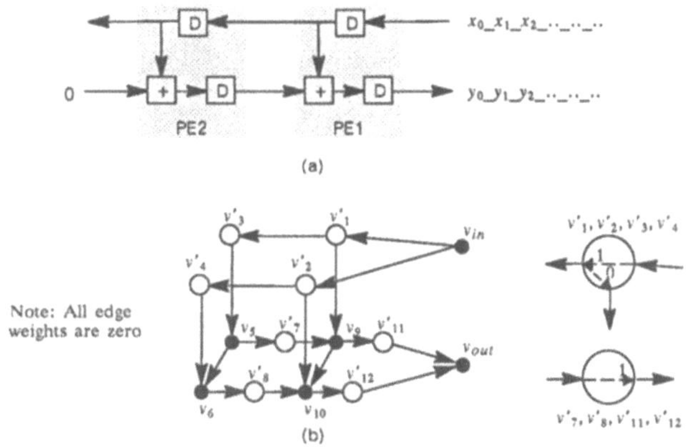

Figure 2: A Systolic FIR Filter (a) Non (two-level) Pipelined Implementation, (b) Graph for (a).

$$k_{in,1} + k_1 + k_{1,3} + k_3 + k_{3,5} + k_{5,6} - k_{4,6} - k_4 - k_{2,4} - k_2 - k_{in,2} = 0$$
$$k_{1,3} + k_3 + k_{3,5} + k_{5,7} + k_7 + k_{7,9} - k_{1,9} = 2$$
$$k_{2,4} + k_4 + k_{4,6} + k_{6,8} + k_8 + k_{8,10} - k_{2,10} = 2$$
$$k_{5,6} + k_{6,8} + k_8 + k_{8,10} - k_{9,10} - k_{7,9} - k_7 - k_{5,7} = 0$$
$$k_{9,10} + k_{10,12} + k_{12} + k_{12,out} - k_{11,out} - k_{11} - k_{9,11} = 0$$

These equations have been obtained by taking the difference of latency through two paths between node pairs $(v_{in}, v_6), (v_1, v_9), (v_2, v_{10}), (v_5, v_{10})$ and (v_9, v_{out}) respectively. Any other equation obtained by other paths can be expressed as a linear combination of the above 5 equations. Now,

$$k_i = 1, \quad \text{for } i = 1,2,3,4,7,8,11,12$$
$$k_{i,j} = \begin{cases} 1 & \text{for } (i,j) = (v_{in}, v_2), (v_{11}, v_{vout}), (v_9, v_{10}), (v_5, v_6) \\ 0 & \text{otherwise} \end{cases}$$

is a feasible solution. This means we have to place pipeline registers to the carry path of each adder and edges $(v_{in}, v_2), (v_{11}, v_{out})$ and keep the latencies of the storage nodes unchanged to get a two-level pipelined design for the systolic FIR filter. This pipelined design can be operated twice as fast as the original design. In general, such a pipelined design can be made to operate m times faster than the original design when an m bit adder is used. The area overhead will be insignificant compared to speed improvements as the pipeline registers need very small silicon area.

4 A Study on Pipelinability

In this section, we study the pipelinability of SAG engines reported in literature [1], [2], [3], [4]. Due to regularity of the graph, only part of the graph will be drawn for the

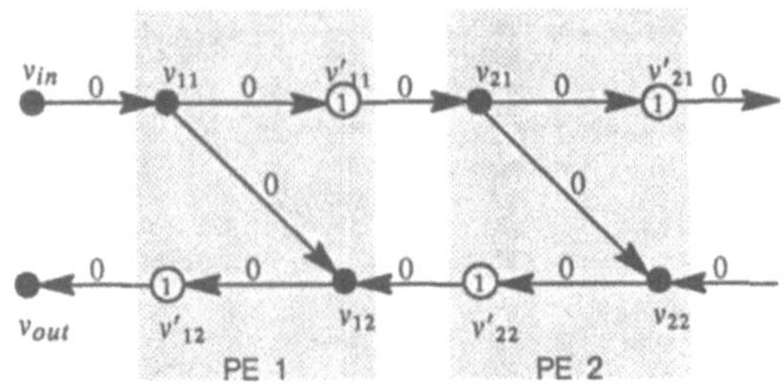

Figure 3: A Partial Graph for Contra-flow Architecture.

discussions in the next sub-sections. Furthermore, we simply omit unnecessary vertex weights.

4.1 Contra-flow vs. Contra-flow Free Architectures

First, let us study the contra-flow architecture given in [1] where the pixel stream travels in opposite direction to instruction and data streams. The relevant partial graph is depicted in Figure 3, where nodes $v_{11}, v_{12}, v'_{11}, v'_{12}$ are nodes in the 1st PE and $v_{21}, v_{22}, v'_{21}, v'_{22}$ are nodes in the 2nd PE. Let $v_{11}, v'_{11}, v_{21}, v'_{21}$ and $v'_{12}, v_{12}, v'_{22}, v_{22}$ be on the instruction and pixel streams respectively. Nodes v_{in} and v_{out} represent the instruction source and display respectively. Let us consider the latencies between nodes v_{11} and v_{12}. Latencies through paths $v_{11}v_{12}$ and $v_{11}v'_{11}v_{21}v_{22}v'_{22}v_{12}$ are 0 and 2 cycles respectively. Now let us assume k_1, k_1, k_2 and k_2 pipeline registers are placed between (v_{11}, v'_{11}), (v_{21}, v'_{21}), (v_{12}, v'_{12}) and (v_{22}, v'_{22}) to get a regular design[4]. Now latencies through paths $v_{11}v_{12}$ and $v_{11}v'_{11}v_{21}v_{22}v'_{22}v_{12}$ are 0 and $k_1 + k_2 + 2$ cycles respectively. Therefoie in order to satisfy the two-level pipelining theorem, $k_1 + k_2 + 2 = 2$. When $k_1 \geq 1$ we get an unrealizable design. Hence we conclude that an SAG engine having a contra flow architecture is not two-level pipelinable[5].

Now let us turn our attention to the contra-flow free architecture presented in [3]. The partial graph is depicted in Figure 4, where nodes v_{11}, v_{12}, v'_{11} are nodes in the 1st PE and v_{21}, v_{22}, v'_{21} are nodes in the 2nd PE. Let $v_{11}, v'_{11}, v_{21}, v'_{21}$ and v_{12}, v_{22} be on the instruction and pixel streams respectively. Nodes v_{in} and v_{out} represent the instruction source and display respectively. Let us consider the latencies between nodes v_{11} and v_{22}. Latencies through paths $v_{11}v_{12}v_{22}$ and $v_{11}v'_{11}v_{21}v_{22}$ are 0 and 1 cycles respectively. Now let us assume k_1, k_1, k_2 and k_2 pipeline registers are placed between (v_{11}, v'_{11}), (v_{21}, v'_{21}), (v_{12}, v_{22}) and (v_{22}, v_{out}) to get a regular design. Now latencies through paths $v_{11}v_{12}v_{22}$ and $v_{11}v'_{11}v_{21}v_{22}$ are k_2 and $k_1 + 1$ cycles respectively. Now, in order to satisfy the two-level pipelining theorem we must have $k_1 - k_2 + 1 = 1$. As the condition $k_1 - k_2 + 1 = 1$ can be satisfied, we can conclude that the direction of the information flow in the contra-flow free architecture does not impose any constrains for two-level pipelining. But we can

[4]In the next section, the necessity for pipeline registers in the instruction streams will become clear.

[5]Note that the systolic FIR filter in our example also has a contra flow architecture, but we were able to design a two-level pipelined systolic FIR filter.

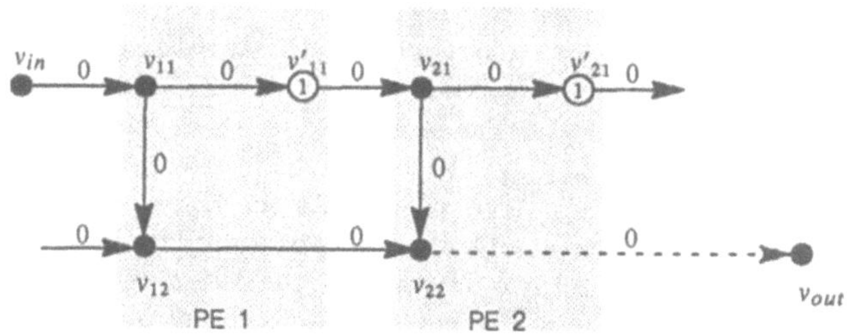

Figure 4: A Partial Graph for Contra-flow Free Architecture.

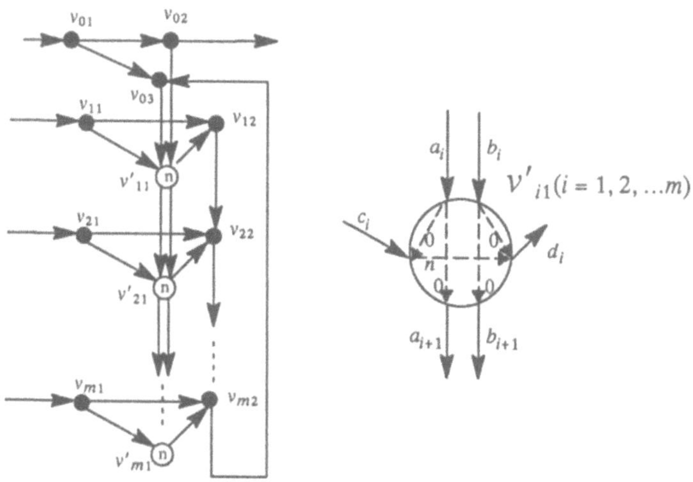

Figure 5: A Partial Graph for a PE Capable of Hidden Surface Removal.

not guarantee that contra-flow free architectures as two-level pipelinable, because in this study we have neglected the internal details of a PE to a great extent. Hence, we classify the contra-flow free SAG engines as potential designs for two-level pipelining.

4.2 Different Instruction Sets

In this sub-section let us study the impact of different instruction sets. First, let us consider the instruction set proposed in [3]. This instruction set is capable of Gouraud shading with hidden surface removal by the Z-buffer algorithm. Let $v_{01}, v_{02}, \ldots, v_{m2}$ be nodes on a PE, see Figure 5, such that nodes $v_{12}, v_{22}, \ldots, v_{m2}$ make the comparison between the current Z value and stored Z value for hidden surface removal. Nodes $v'_{11}, v'_{21}, \ldots, v'_{m1}$ store the Z value to be used with the next instruction according to the results of the comparison. Nodes v_{01}, v_{02}, v_{03} are on the instruction stream such that v_{02} and v_{03} generate the write and read signals for the depth storage. The graph has been drawn by assuming that the depth is calculated by m bit fixed-point numbers and the instructions for hidden surface removal are issued at least n cycles apart as done in [3]. Note that, in this graph there are several loops passing through the critical path. (We have assumed that the critical path lies on the comparison circuit). Application of the two-level pipelining theorem reveals that pipeline registers can be inserted into the critical path if we reduce the latencies of the nodes $v'_{11}, v'_{21}, ..., v'_{m1}$. The reductions in storage latencies are necessary due to directed loops such as $v'_{11}v_{21}v_{22}...v_{m2}v_{03}$. As the nodes $v'_{11}, v'_{21}, ..., v'_{m1}$ represent storage logic in the design[6], the comparator unit can only be pipelined into maximum depth of $n-1$. In general $m \gg n$ ($m = 32$, $n = 4$ for the design in [3]) and hence the speed improvement is limited. When $n = 1$, the comparator can not be two-level pipelined.

The instruction set presented in [4] has been proven to be two-level pipelinable to any depth. Due to the complexity of the graph the proof is omited and in the next section we will present a two-level pipelined architecture.

5 A Two-level Pipelined SAG Engine

First let us study a non two-level pipelined SAG engine capable of executing the instruction set given in [4] (see Table 1). Figure 6 shows a non-pipelined architecture of a PE which can execute the above instruction set. The registers A,B and C are used to store intensity, its first derivative and its second derivative for interpolations. The register P holds the accumulated intensity (i.e. local pixel storage), which is sent to the output driver and then cleared by the 'REF' instruction. The 'SETI', 'SETDI', 'SETDDI' and 'DIS' instructions disable the registers from being over-written by the next 'EVAL*' instruction. The multiplexers at the right side are used to handle some functional discontinuities. For example, when the PE is (not) within X and X+DX, interpolation is (not) performed. Hence, the right multiplexer on the data stream selects the correct data to the output according to the location of the PE. Furthermore, at some PE locations the state of the instruction is changed. For example, at PE locations X and X+DX the 'EVAL*' instructions are activated and converted into a 'NOP' instruction respectively. This is achieved by the multiplexer on the instruction stream. The discontinuities of the operations performed by each PE depend on the value of X, DX and instruction type. Therefore, the control signals for the multiplexers are generated in conjunction with the carry signals (Cx in Figure 6) from the adder and the instruction in the processor.

The target architecture in Figure 6 can be converted into two-level pipelined design as outlined below. First the graph of the architecture is constructed. Then the critical paths are identified. In our target architecture, the critical paths pass through the carry ripple path in the adder. The places where the pipeline registers must be placed in the critical paths are then decided to meet the given speed requirements. Then locations of other pipeline registers are determined by solving the linearly independent equations obtained by the application of the two-level pipelining theorem. The graph of the architecture is not given due to its complexity and hence more details of the steps we outlined above are

[6]The storage nodes must store data at least for one cycle, otherwise we convert the synchronous design into an asynchronous design.

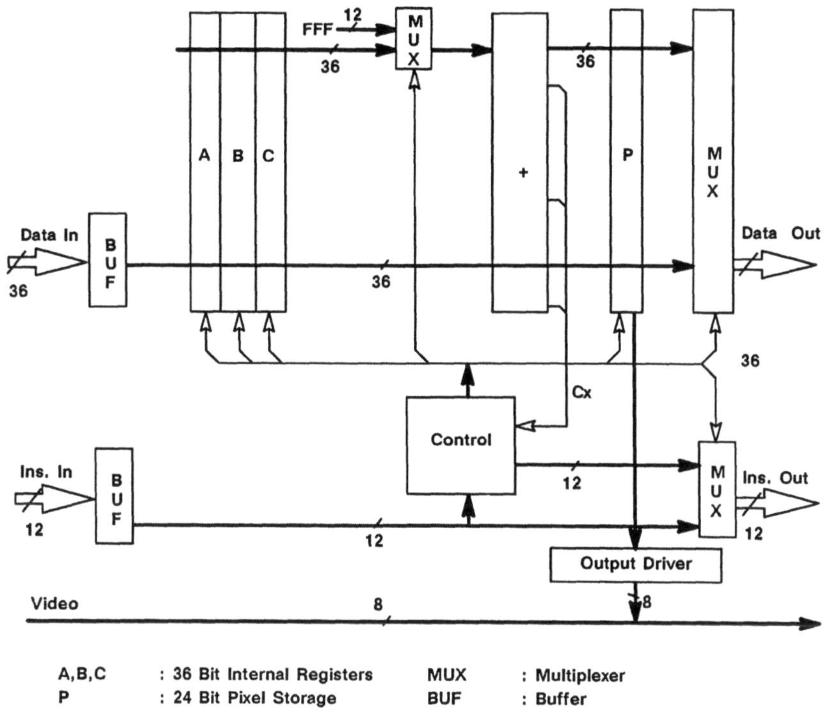

Figure 6: Architecture of a Non (two-level) Pipelined SAG Engine for the Instruction Set in [4].

also omited. A two-level pipelined architecture derived by this approach is depicted in Figure 7. Though the architecture in Figure 7 is pipelined to depth 2, it can be pipelined to the bit-level.

Some explanations are given in this paragraph to understand the operation of the pipelined engine. As the critical paths of this circuit lie through the adder, 2 pipeline registers have been placed on the carry path. As control signals for the multiplexers on the right side are generated in conjunction with the carry signals from the adder, the (data) inputs to these multiplexers are delayed by pipeline registers, which can be seen in Figure 7. As the carry signal from the low order bits to higher order bits travels in different time slots, the input data to the adder must also be delayed and hence pipeline registers are placed on the control signals of registers A,B,C,D and multiplexer in the middle. Furthermore, the latency of the carry introduces latency in the adder output. Due to this reason, pipeline registers are introduced on the right hand multiplexer control signals. Pipeline registers on the pixel stream and on the control signals of the output driver are also placed due to similar reasons.

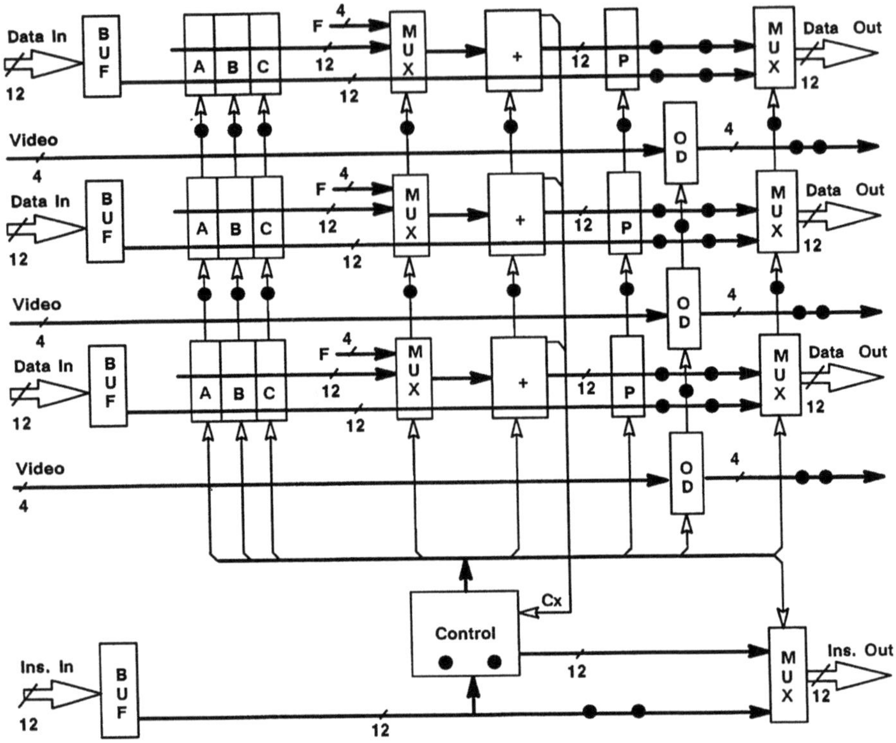

A,B,C	: 36 Bit Internal Registers	MUX	: Multiplexer
P	: 24 Bit Pixel Storage	BUF	: Buffer
OD	: Output Driver	●	: Pipeline Register

Figure 7: A Two-level Pipelined Architecture.

6 Conclusions

The operating speed of the systolic array is determined by the most complex operation performed. Hence the maximum operating speed of an SAG engine tends to reduce as more complex operations are introduced to improve the image quality. Furthermore, high resolution displays also demand high speed operation. In order to improve the speed of SAG engines, the use of pipelined functional units were proposed. Even though pipelined functional units need more hardware for pipeline registers, we think two-level pipelining of SAG engines will be very attractive for tomorrow's computer graphics hardware due to the following major reasons. First, SAG engines generate the pixels in real-time from an object representation and hence one can improve the interaction speed by replacing the conventional frame buffer by a structured object list as proposed in [12]. The SAG engine based systems are more compact than systems based on processor-per-pixel or processor-per-object providing the same functionality, image quality and interaction speed. As the scaling of minimum feature sizes and wafer scale integration increase the device count dramatically, one has very feasible means to provide the additional hardware for two-level pipelining of SAG engines.

A graph-theoretic methodology was presented to transform SAG engines built from non-pipelined functional units into two-level pipelined designs. In an SAG engine, the number of PEs could be large ($256 \sim 4000$) and hence the number of nodes and edges in the graph could be very large ($10,000 \sim 100,000$). Hence, the execution time of the re-timing procedure on a computer will also be very large. Currently we are looking into fast implementations of re-timing by exploiting the regularities in the graph.

Acknowledgments

Thanks to A.A.M. Kuijk and E. Blake at Interactive Systems Group, CWI, Amsterdam for helpful discussions. Thanks also to F. Moelart El-Hadidy, G. Karagiannis and S. Gerez at Laboratory of Network Theory, Twente University, Enschede for proof reading and helpful discussions.

References

[1] Nader Gharachorloo and Christopher Pottle, *SUPER BUFFER: A Systolic VLSI Graphics Engine for Real Time Raster Image Generation*, Proceedings of 1985 Chapel Hill Conference on VLSI, Computer Science Press, 1985, pp. 285-305.

[2] Nader Gharachorloo, Satish Gupta, Erdem Hokenek, Peruvemba Balasubramanian, Bill Bogholtz, Christian Mathieu and Christos Zoulas, *Subnanosecond Pixel Rendering with Million Transistor Chips,* Proceedings of SIGGRAPH 88, August 1988, pp. 41-49.

[3] Teiji Nishizawa, Takeru Ohgi, Kazuyasu Nagatomi, Hiroshi Kamiyama and Kiyashi Maenobu, *A Hidden Surface Processor for 3-Dimension Graphics*, ISSCC 88, February 1988, pp. 166-167.

[4] J.A.K.S. Jayasinghe, A.A.M. Kuijk and L. Spaanenburg, *A Display Controller for a Structured Frame Store System*, in Advances in Computer Graphics Hardware III, Springer-Verlag, 1989.

[5] H. Fuchs and J. Poulton *Pixel-planes: A VLSI Oriented Design for a Raster Graphics Engine*, VLSI Design, Vol. 3, No. 3, 1981, pp. 20-28.

[6] Bengt-Olaf Schneider, *PROOF: An architecture for rendering in Object Space*, in Advances in Computer Graphics Hardware III, Springer-Verlag, 1989.

[7] D.A. Hodges and H.G. Jackson, *Analysis and Design of Digital Integrated Circuits*, McGraw-Hill, Inc. 1983.

[8] H.T. Kung and M.S. Lam, *Fault-Tolerance and Two-level Pipelining in VLSI Systolic Arrays*, 1984 Conference on Advance Research in VLSI, M.I.T., 1984, pp. 74-83.

[9] C. E. Leiserson and J.B. Saxe, *Optimizing Synchronous Systems*, Journal of VLSI and Computer Systems, Vol.1, No.1, 1983, pp. 41-68.

[10] S.Y. Kung, *VLSI Array Processors*, Prentice Hall, 1988.

[11] H.T. Kung, L.M. Ruane and D.W.L. Yen, *Two-level Pipelined Systolic Array for Multi-dimensional Convolution*, Image and Vision Computing, Vol.1, No. 1, February 1983, pp. 30-36.

[12] P.J.W. ten Hagen, A.A.M. Kuijk and T. Triekens, *Display Architecture for VLSI-based Graphics Workstation*, in Advances in Computer Graphics Hardware I, Springer-Verlag, 1987.

[13] J.A.K.S. Jaysinghe, *Speed Optimization of VLSI/WSI Array Processors*, Internal report under preparation.

Part 2

Systems

A Dedicated Graphics Processor SIGHT-2

Masaharu Yoshida, Tadashi Naruse and Tokiichiro Takahashi

Autonomous Robot Systems Laboratory
NTT Human Interface Laboratories
NIPPON TELEGRAPH AND TELEPHONE CORPORATION
1-2356 Take Yokosuka-Shi
Kanagawa 238-03 Japan

Abstract

SIGHT-2 is a multiprocessor system that is intended to efficiently execute the ray tracing algorithm. To achieve high efficiency, three kinds of parallel execution mechanisms; (i) a multiprocessor configuration, (ii) a parallel execution of three dimensional vector operations, and (iii) functionally distributed parallel processing are introduced. Owing to the latter two techniques, each processing element (PE) has the ability to execute the standard ray tracing algorithm 10 times faster than a VAX11/780 with a floating point accelerator. In the present configuration, SIGHT-2 utilizes 16 PEs, which results in a peak power of 66.72 MFLOPS / 133.28 MIPS. During ray tracing, the efficiency of each PE is over 99% under static load balancing.

In this paper, SIGHT-2 system architecture, its PE configuration, and VLSIs design are discussed. The system performance is also discussed.

1. Introduction

With computer graphics, computers are able to convert large amounts of complex data into visual forms that are more readily understood by humans. There is, however, a tradeoff between visual detail and the computation cost needed to render the image. Among the various computer graphic techniques, the ray tracing algorithm offers the highest accuracy and realism; both important characteristics for image comprehension[1].

The ray tracing algorithm duplicates the action of human vision but in reverse. Instead of light rays being collected by an eye, ray-object intersections are modeled from the eye point through each pixel on the image plane to the objects forming the image. The ray tracing algorithm can yield extremely detailed images but at a very high computation cost. This cost can be reduced by increasing the processor speed or the number of processors.

The basic ray tracing algorithm is well suited to parallel processing because there is no need for inter-PE communication. Since the ray tracing algorithm is extremely time consuming, many multiprocessor systems have been developed (for example, [2]). While these multiprocessor systems do decrease image generation time, the achievable speeds are still too slow for practical commercial use.

By analyzing the ray tracing algorithm in detail, we have clarified that it can be performed in the following three levels of parallel execution[3,4].

(1) Pixel level parallel execution :
each pixel value can be calculated independently, since rays do not interact with each other.

(2) Operation level parallel execution :
three dimensional vector operations can be executed in parallel with respect to the x, y, and z axes. They occupy the largest part of the ray tracing algorithm[3].

(3) Function level parallel execution :
three function level processes can be executed in parallel: floating point calculations like intersection calculations, integer operations like address calculation, and data transfer.

Based on the above analysis, we have proposed a ray tracing computer architecture, SIGHT[3], and developed a SIGHT prototype system with one PE[3]. Its PE employs special hardware to perform three dimensional vector operations in parallel. This innovation makes it possible to execute the ray tracing algorithm very quickly.

The SIGHT prototype system verified that its PE architecture is efficiently suited to the parallel execution of the three dimensional vector operations. From SIGHT, we developed the SIGHT-2 prototype system which is a multiprocessor system with 16 PEs.

In this paper, first, the parallel execution method of the ray tracing algorithm and the architecture of SIGHT will be recapitulated. Second, the SIGHT-2 PE and VLSI designs are detailed. Finally, measured performance is compared to that of a VAX11/780.

2. Parallel Execution of Ray Tracing Algorithm

In this section, we discuss the parallel execution of the ray tracing algorithm. First, the basic ray tracing algorithm is given, second, the pixel level parallel execution method is explained, and finally, the operation level parallel execution method is detailed.

2.1 Ray Tracing Algorithm

The basic concept of the ray tracing algorithm is that the computer simulates a pin hole camera. That is, the algorithm generates the image on the screen by tracing rays from the eye to the objects as shown in Figure 1.

The basic image generation steps are as follows :

(1) The ray equation is calculated from the eye position to a pixel position.

(2) The intersections of the ray with each object are calculated.

(3) If the object closest to the screen is transparent, the reflected and refracted ray equations are calculated, and the new rays are recursively traced with the same steps as in (1) and (2). If the closest object is a mirror, the reflected ray is traced in the same way as above. if the closest object is another type, then step (4) is executed. If no object intersects with the ray, then step (4) is executed.

(4) The shading calculation is executed.

The above calculation steps are repeated until all pixel values of the screen and screen image are generated.

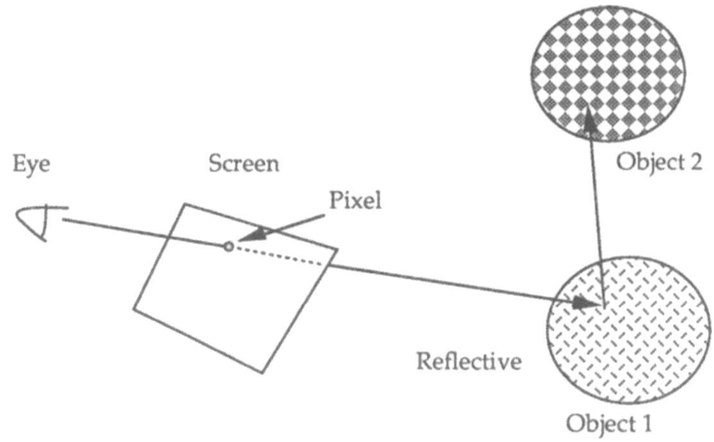

Figure 1 Ray Tracing Algorithm

The calculation cost of the ray tracing algorithm is proportional to the product of the number of pixels and objects. Therefore, the ray tracing algorithm needs enormous computing time to generate a photo realistic image, and parallel execution is necessary to realize high speed computation. In the following subsections, we discuss the parallel execution of the ray tracing algorithm.

2.2 Pixel Level Parallel Processing

In the ray tracing algorithm, since there is no interference among rays, each pixel value on the screen is calculated independently. This means that simple multiprocessor architectures are well suited to the ray tracing algorithm. If each processing element (PE) holds a copy of the entire data and program codes that are needed to generate the desired image, there is no inter-PE communication. We call this kind of parallel execution, "Pixel level parallel processing".

There have been some systems developed from this idea [for example, 2]. While these multiprocessor systems reduce image generation time considerably, the achievable speeds are still insufficient for practical use. For this reason, we can not expect sufficient performance with only pixel level parallel processing.

2.3 Operation Level Parallel Processing

In the previous papers[3,4], we have shown another kind of parallelism for the ray tracing algorithm. This is summarized in this subsection.

As T. Whitted pointed out[1], intersection calculations occupy 60% ~ 95% of the total computation time of the ray tracing algorithm. So, let us consider the intersection calculation.

The ray equation is given by.

$$\mathbf{x} = \mathbf{a} * t + \mathbf{v} \quad (1)$$

where, **V** is the eye position vector, **a** is the direction cosine vector, and **t** is a parameter.

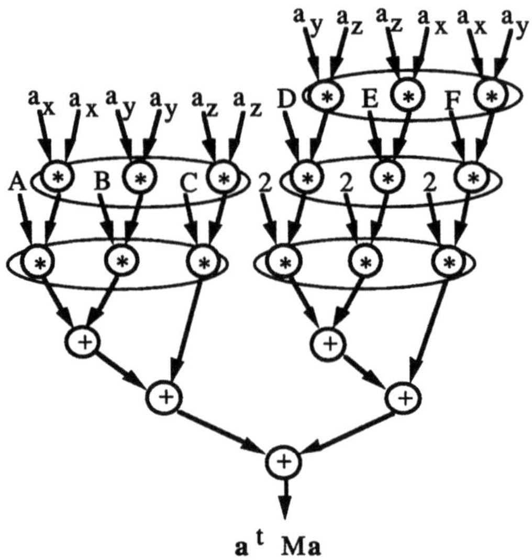

Figure 2 The Data Flow Graph

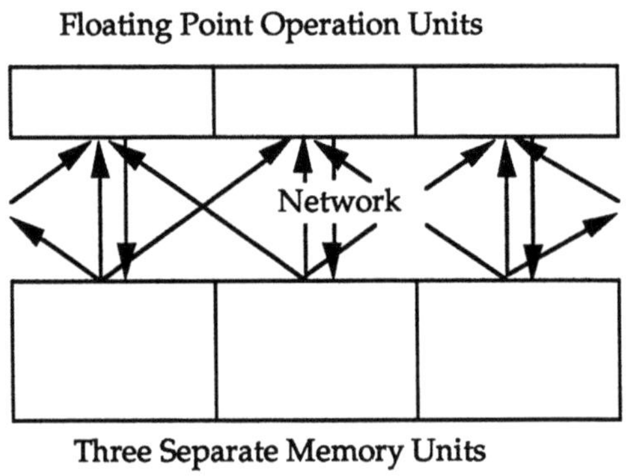

Figure 3 Three Dimensional Vector Parallel Execution Mechanism

In the case of a quadric surface, the object is given by the quadratic form.

$$\mathbf{x}^t \mathrm{M}\, \mathbf{x} = 1 \quad (2)$$

$$\mathrm{M} = \begin{pmatrix} A & F & E \\ F & B & D \\ E & D & C \end{pmatrix}$$

where, M is a coefficient matrix. By substituting equation (1) into equation (2),

$$\mathbf{a}^t \mathrm{M}\, \mathbf{a} * t^2 + 2\, \mathbf{a}^t \mathrm{M}\, \mathbf{v} * t + \mathbf{v}^t \mathrm{M}\, \mathbf{v} = 1 \quad (3)$$

is obtained.

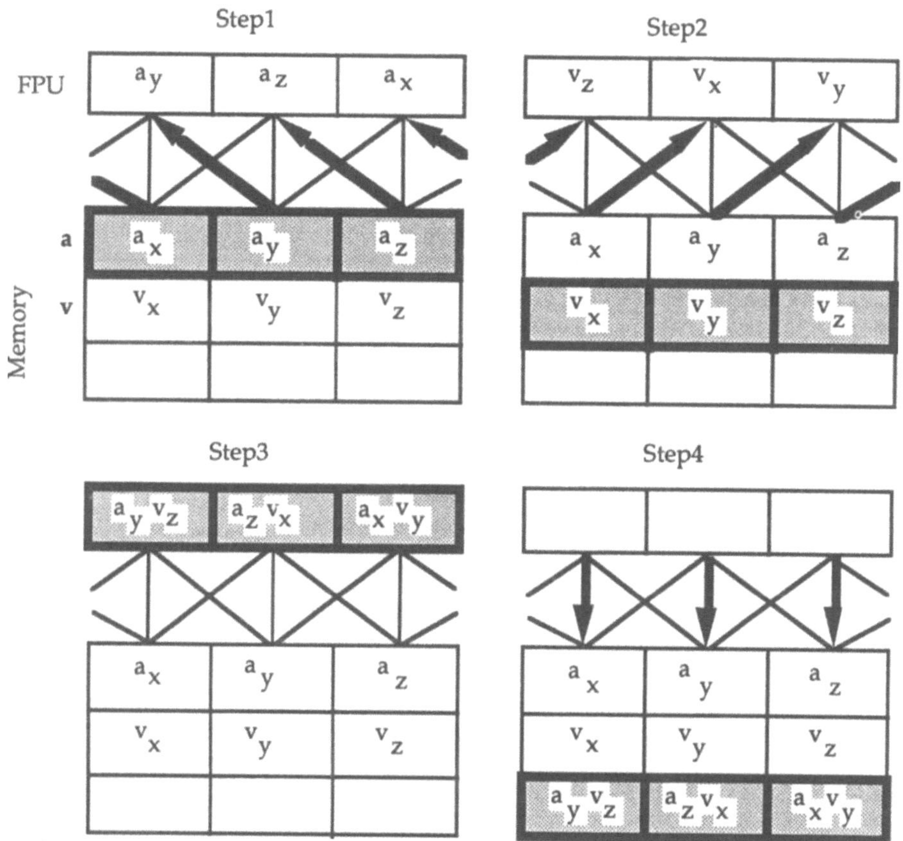

Figure 4
Example of Parallel Three Dimensional Vector Operations in the TARAI Unit

By solving equation (3), the intersection is obtained. In equation (3), it takes much time to calculate the coefficients of the quadratic equation.

Figure 2 shows a data flow graph that computes a coefficient: $\mathbf{a}^t M \mathbf{a}$. From this graph, it is shown that the input data of three multiplication operations enclosed by ellipse are regular and symmetric with respect to the x, y, and z axes. This means that these multiplications can be executed efficiently in parallel by using three Floating Point operation Units (FPUs) and three separate memory units connected by a simple network. This concept was demonstrated to be correct by the tests performed on the TARAI unit in 1987[3].

One realization of the execution mechanism is shown in Figure 3. How the above operations are executed on this mechanism will be explained. Let us consider three products

$$a_y * v_z,\ a_z * v_x,\ a_x * v_y.$$

We store the data in the memories as shown in Figure 4, i.e. data for the x-axis is stored in the left memory, data for the y-axis in the center memory, and data for the z-axis in the right memory. Thus multiplication operations can be executed as follows,

(1) a_x, a_y, a_z are transferred to the upper-left operation units (step 1),
(2) v_x, v_y, v_z are transferred to the upper-right units (step 2),
(3) multiplications are executed (step 3),
(4) results are transferred to lower memories (step 4).

It should be noted that no memory access conflict occurs during the execution of these operations.

Since most of the three dimensional vector operations have regularity and symmetry, this parallel execution mechanism works well for three dimensional vector operations. We call this unit, "TARAI", and the parallel process, "Operation level parallel processing".

3. SIGHT Architecture

SIGHT employs three kinds of parallel execution mechanisms;
(1) Pixel level parallel processing,
(2) Operation level parallel processing,
(3) Functionally distributed parallel processing.

The first two were described in the previous section. The last process will be explained in Section 3.3.

3.1 SIGHT System Architecture

SIGHT is a multiprocessor system for the fast execution of the ray tracing algorithm. SIGHT-2 was designed using the SIGHT architecture.

The SIGHT-2 prototype system is shown in Figure 5. There are 16 PEs, a host interface, a data collector and a frame buffer. Each PE has six communication interfaces; one is to the host processor, one is to the frame buffer through the data collector, and four are to the neighboring PEs.

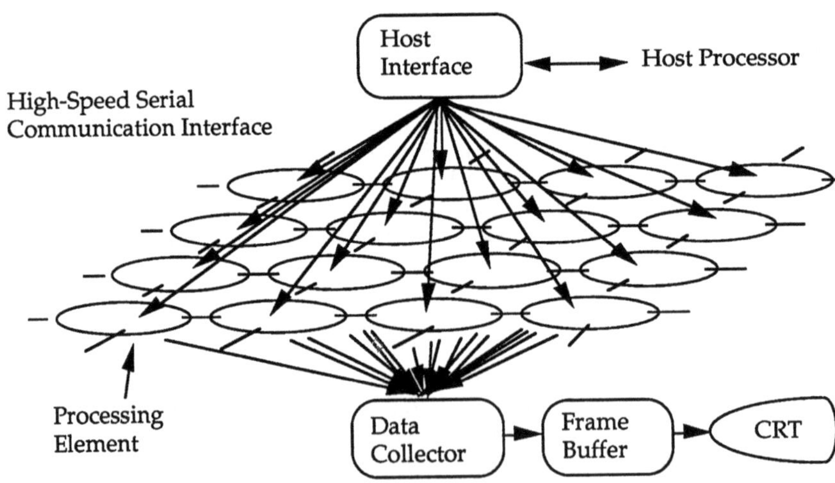

Figure 5 SIGHT-2 System Configuration

Inter-PE communication lines are prepared for the space subdivision ray tracing algorithm that will be implemented at a later date. Each communication line is a high-speed serial line, because it is necessary to reduce the connections for effective VLSI design. Data is packetized and transmitted through these communication lines.

The ray tracing program code and data for image creation are supplied by the host processor. Each PE executes the ray tracing algorithm and sends the pixel values to the frame buffer through the data collector. Finally, the images are displayed on the CRT.

3.2 The SIGHT PE Architecture

The PE architecture of SIGHT is shown in Figure 6. There are four major units; TARAI, MP, Direct Memory Access (DMA), and Data Base Memory (DBM). In the TARAI unit, there are three floating point operation units, three register files, and a TARAI network connecting FPUs to register files. The TARAI unit executes three dimensional vector operations in parallel. The MP unit contains an integer ALU, a register file and six channel high-speed serial communication interfaces. The MP unit controls the TARAI and DMA units, and communicates with the host processor, the frame buffer and four neighboring PEs. The DMA unit transfers the data between the register files and the DBM unit. The DBM unit is the main memory and holds the entire set of object and mapping data needed by each PE.

TARAI and MP units can not access the DBM unit directly. They can access only the register files. The register files and DMA units act as a cache memory controlled by the MP unit. This configuration makes it possible to construct a high-speed, high-density and large capacity memory system by using two types of memory chips: high speed, small capacity SRAMs used to construct register files; and low speed, high capacity DRAMs for the DBM unit.

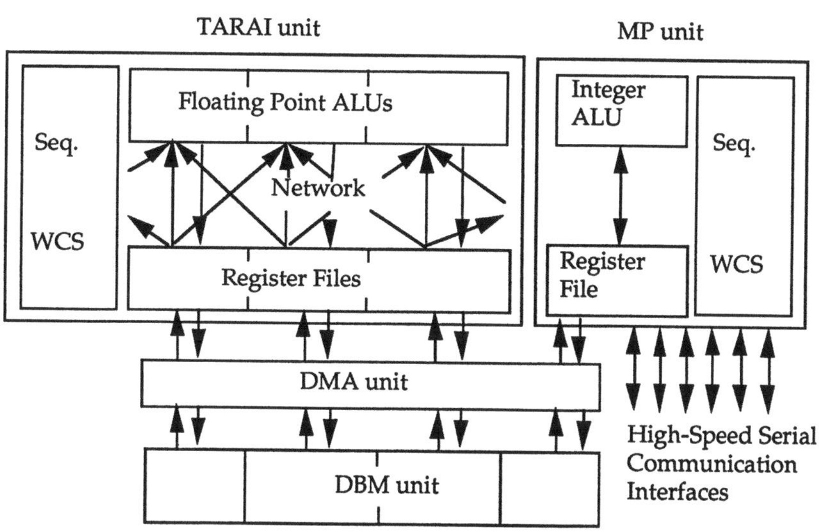

Figure 6 Processing Element (PE)

As the TARAI and MP units are controlled by separate microprogram both units can operate independently. The ray tracing program is written in micro code.

In the SIGHT PEs, there is only one dedicated hardware mechanism to execute the three dimensional vector operations in parallel. The rest of the ray tracing algorithm is executed as a microprogram. We consider this configuration is important because the ray tracing algorithm is still being optimized and there are many possibilities of improving it.

3.3 Functionally Distributed Parallel Processing

The most time consuming component of the ray tracing program is the calculation of ray-object intersections. Sequential execution of ray-object intersection is shown in Figure 7, and is summarized as follows.

(1) The MP unit calculates the stored address of one object in the DBM unit.

(2) The MP unit invokes the DMA unit.

(3) The DMA unit transfers the object definition data from the DBM unit to the register files.

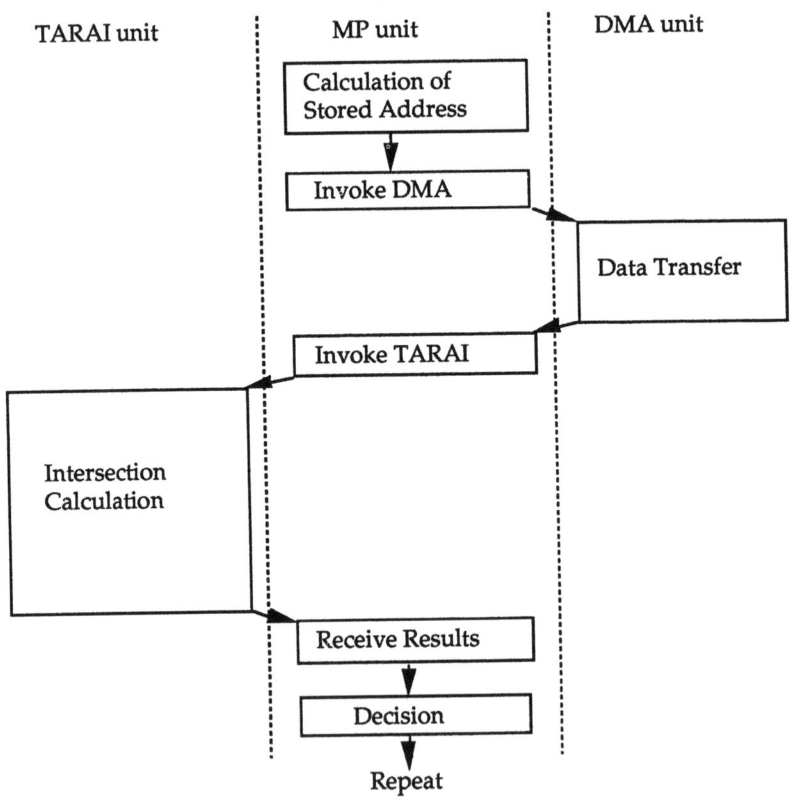

Figure 7 Execution of Ray Tracing

(4) After finishing data transfer, the MP unit invokes the TARAI unit.
(5) The TARAI unit executes the intersection calculation.
(6) The MP unit receives the intersection data from the TARAI unit, compares it with the previously calculated data, selects the object closer to the screen, and memorizes it.
(7) If there is more data to calculate, repeat above steps for a new object.

Since the TARAI unit, MP unit and DMA unit can operate concurrently, parallel pipelined operation is possible as shown in Figure 8. While the TARAI unit calculates intersections, the MP and DMA units prepare the next object's data, and the MP unit checks the previous result of the intersection calculation by the TARAI unit. We call this operation, "Functionally distributed parallel processing".

4. SIGHT-2 System Implementation

In this section, the SIGHT-2 system hardware configuration will be explained. After explaining PE configuration, we discuss VLSI design considerations for the PEs employed in SIGHT-2.

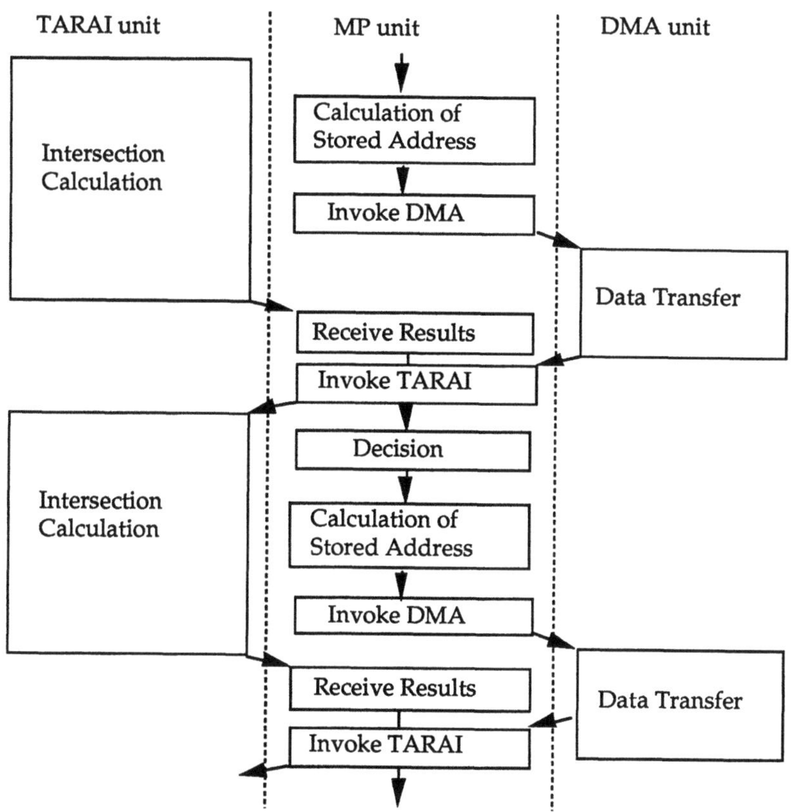

Figure 8 Pipelined Execution of Ray Tracing

4.1 PE Configuration

The hardware configuration of a PE is shown in Figure 9. The TARAI Controllers and Master Controller are custom VLSIs, which were designed for SIGHT-2. Their details are described in Section 4.3.

TARAI unit

The TARAI unit is made up of three FPUs and three register files. The TARAI network is constructed with three time-shared buses, and it will be described later. Each FPU contains a floating point ALU, a floating point multiplier, ROM function tables, and a TARAI Controller VLSI. Frequently referenced functions, such as square root, sine, cosine, and gaussian, are provided in the function tables for the high speed calculation.

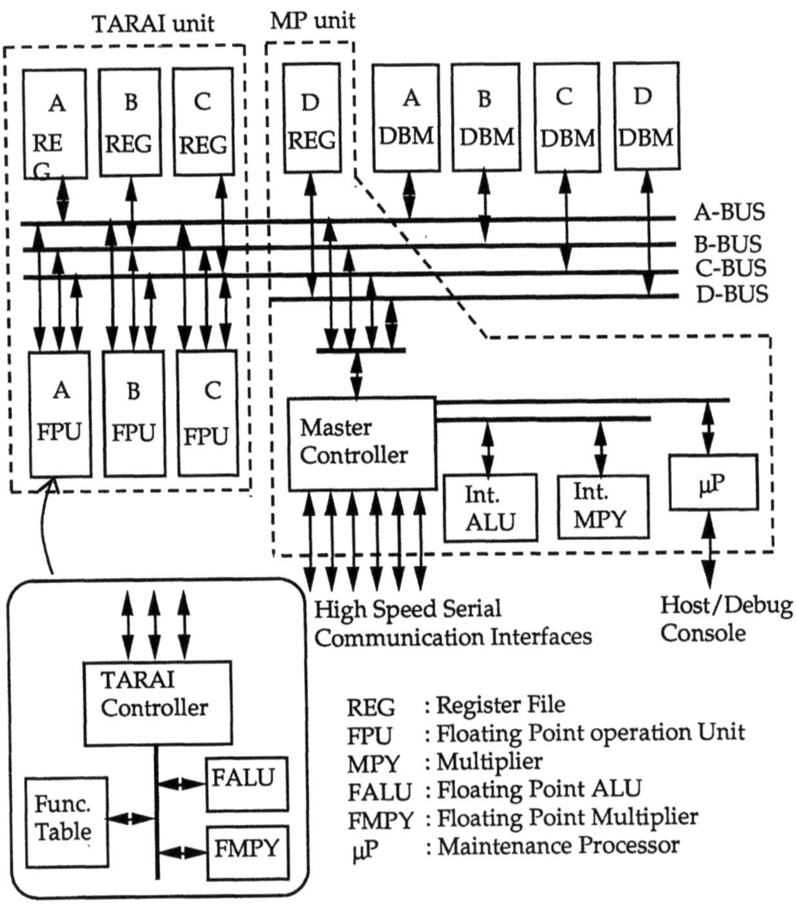

Figure 9 The hardware configuration of a PE

The TARAI Controller VLSI, controlled by the microprogram of the TARAI unit, controls the bus selection of each FPU. The bus architecture reduces the number of pins in the TARAI Controller VLSI chip.

MP unit

The MP unit contains a 32 bit integer ALU (Int. ALU) and multiplier (Int. MPY), a maintenance processor (μP) and a Master Controller VLSI. The Master Controller VLSI includes six channel high-speed serial communication interfaces and a DMA unit. The capacity of each serial communication interface is 2 mega bytes per second.

The basic clock rate of the MP unit is 120 nano seconds, which is synchronized to the clock in the TARAI unit. The MP unit can invoke the TARAI unit microprogram routine at any address of the TARAI unit's WCS.

The μP controls the microprogram execution of both the TARAI unit and MP unit. It also supports a microprogramming developing environment by offering facilities such as microprogram loading, inline assembler, and debugging.

Bus

The TARAI network is constructed with three 32 bit time-shared buses; A-BUS, B-BUS, and C-BUS. This architecture realizes the concurrent execution of floating point operations and data transfer between the register files and the MP unit or DBM unit without decreasing performance of the TARAI unit. The three FPUs operate in a three address mode.

The bus timing of the TARAI unit is shown in Figure 10. The bus clock period is 120 nano seconds. For the single precision floating point data format, six time slots are needed. As a result, the TARAI unit cycle time is 720 nano seconds, and the TARAI unit can perform at up to 4.17 MFLOPS in single precision.

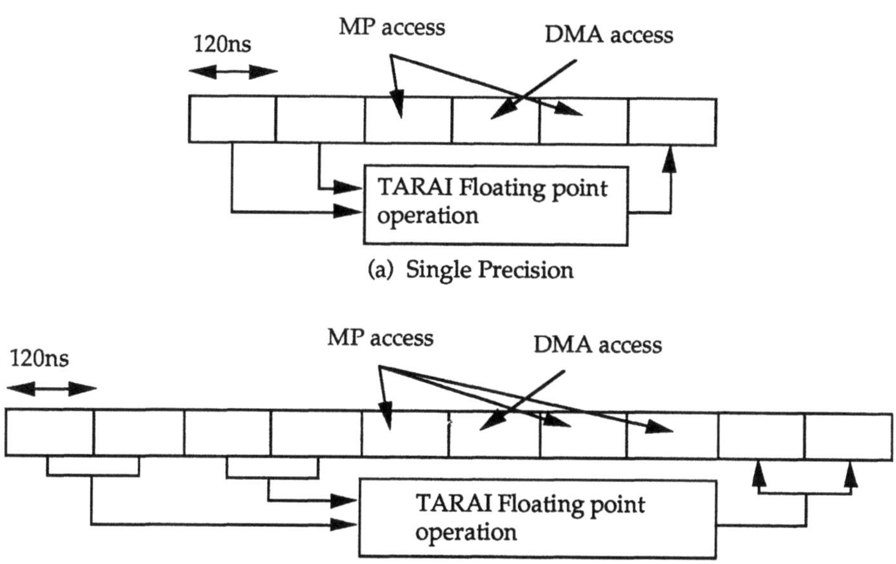

(a) Single Precision

(b) Double Precision

Figure 10 Bus Timing

The first 2 time slots are assigned to transfer two operands from the register files, and the final time slot is to store operation results in the register files. While the FPUs execute floating point operations, they do not access the buses. The remaining time slots are used for the MP and DMA units to access TARAI unit's register files.

The time slots for the TARAI and DMA units are assigned statically, while the time slots for the MP unit are assigned dynamically. The MP unit can use time slots only when other units are not using the buses. The bus is arbitrated by the Master Controller VLSI.

For the double precision floating point data format, ten time slots are needed as shown in Figure 10. Two time slots are needed to transfer one word between the FPUs and the register files, and one extra time slot is needed to execute the double precision floating point operations.

4.2 Design Consideration of VLSI Chips

We developed two types of VLSI chips. Since a fair amount of miscellaneous control gates are included on the VLSI chips, each PE can be implemented on just two printed circuit boards. The key design factors of the VLSIs are as follows :

TARAI Controller

(1) Bus interfaces:
As shown in Figure 9, the TARAI Controller must interface four 32 bit buses; three constitute the TARAI network, and one connects the floating point ALU, floating point multiplier, and function table. About one half of the TARAI Controller chip pins are tied to these buses. These buses are also used for chip testing. Through them, test data can be efficiently read and/or written.

(2) Nanoprogram control:
For the single precision floating point data format, each TARAI instruction is executed in 6 bus time slots (Figure 10). Each FPU is controlled by a nanoprogram. As the architecture of the TARAI unit is not yet optimum, it is desirable to be able to modify the instruction set; moreover, it is necessary to create application specific instruction sets. Therefore, we added RAM storage to the TARAI Controller chip to contain the nanoprogram.

Master Controller

(1) Communications:
Since each PE communicates with the host processor, four neighboring PEs, and the frame buffer, six communication lines are needed. Due to the restriction in pin number, a serial communication interface was adopted. The capacity of each serial communication line is 2 mega bytes per second. This performance is sufficient for host-PE and PE-frame buffer communications. This is because the interface between the host interface and host processor is constructed by using the DR11W[1] interface, the capacity of which is 2 mega bytes per second. In addition, the pixel value execution task takes much longer than communication tasks.

---

[1] Digital Equipment Corporation

Though inter PE communication is not necessary for standard ray tracing, it is provided for implementing advanced ray tracing techniques, such as space sub-division ray tracing. In such cases, the inter PE communication speed is fast enough, because the intersection computation takes much longer than communication tasks.

(2) Bus interfaces:
Due to the restriction in the number of pins, the Master Controller chip interfaces four multiplexed buses (A,B,C,D-Bus). This does not decrease performance, because only one register file is accessed by the MP at a time.

4.3 VLSI Chips

We developed two custom VLSIs: TARAI Controller and Master Controller. Both chips are 1.5 micron CMOS channel-less gate arrays.

TARAI Controller Chip

The block diagram of the TARAI Controller VLSI chip is shown in Figure 11. The TARAI Controller was designed to yield a function sliced architecture for the TARAI network, and three chips are used to construct the network. Each Controller includes a bus coupler, floating point operation unit support circuits, and writable control store (WCS) for the nanoprogram. In addition, a 32 bit integer ALU, different from that in the MP unit, is also included for register file address calculations.

The nanoprogram controls the operation of the floating point ALU, floating point multiplier, and function table. Since the nanoprogram is stored in the WCS, the microprogram architecture of the TARAI unit can be changed to fit various applications. The size of the WCS is 128 words each 48 bits wide. The nanoprogram is upward loaded or downward loaded by the μP through the A-BUS, B-BUS, and C-BUS.

The Controller interfaces four 32 bit bi-directional buses. In general, a VLSI chip with many output pins may generate electric noise when many output signals change simultaneously. This is no problem for the TARAI Controller because only one bus is driven by the TARAI Controller at a time.

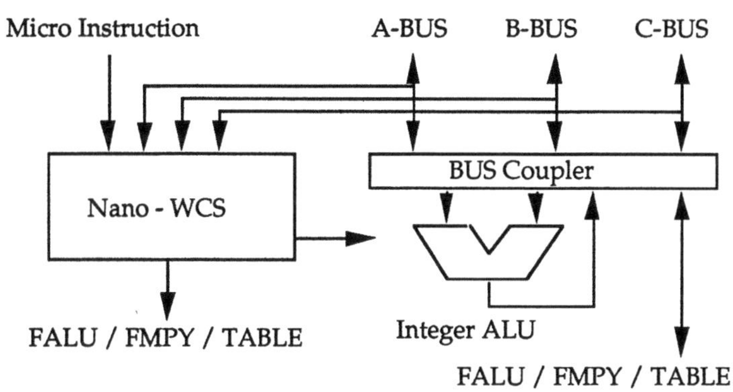

Figure 11 TARAI Controller

The TARAI Controller VLSI chip has 28 kilo gates, housed in a 233 pin Pin Grid Array (PGA) package.

Master Controller Chip
The block diagram of the Master controller VLSI chip is shown in Figure 12. It includes four MP bus interfaces, a 32 bit barrel shifter, six channel high-speed serial communication interfaces, the DMA unit, a μP interface, and a system timing generator. Each MP bus interface contains a 32 bit bi-directional data register and a 16 bit address register/ incrementer. The address increment value is micro-programmable in the range of 0 to 15.

High speed serial lines are provided for the transmission of data packets to neighboring PEs, the host processor, or the frame buffer through the data collector. The length of each packet is 64 bits. Because 11 bits are provided for packet address, it is logically possible to connect over 2,000 PEs. The barrel shifter is included to manipulate the packets.

The system timing generator supplies all timing clocks used in the PE and also generates the bus time slots clock.

The system timing generator controls the execution timing of the TARAI unit , MP unit and DMA unit, arbitrates four buses; A-BUS, B-BUS, C-BUS and D-BUS, and refreshes DRAMs in the DBM unit.

This VLSI has 27 kilo gates in a 299 pin PGA package.

4.4 System Overview

SIGHT-2 Prototype System
The appearance of the SIGHT-2 prototype system is shown in Figure 13. The front door has been removed to take this picture. The cabinet is about 175 cm high, 130 cm wide, 110 cm deep. 16 PEs are vertically mounted in the upper half of the cabinet. The host interface, the data collector and the frame buffer are mounted in the lower left of the cabinet.

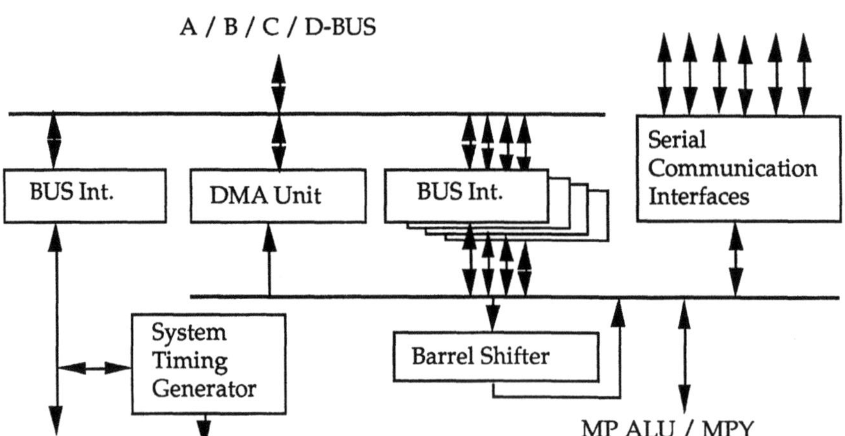

Figure 12 Master Controller

The ribbon cables are the high-speed serial communication lines. Each PE communicates with its neighbors, the host processor and the data collector through the serial lines. There is no mother board (back plane) to make it easy to change connections between PEs.

PE of SIGHT-2

A PE of SIGHT-2 is shown in Figure 14. Each PE is consists of two piggybacked printed circuit boards. The board size is 53 cm wide, 47 cm deep.

A SIGHT prototype with a single PE was developed and tested in 1987[3]. The single PE was implemented on 15 wire wrapped circuit boards. In SIGHT-2 the number of boards is reduced to only two. This compact implementation is made possible with the two specialized VLSIs.

Figure 13
The SIGHT-2 Prototype System

Figure 14 PE

The upper board is shown in Figure 15. The majority of this board is occupied by memory chips. The upper center area is the DBM unit, which consists of 64 mega bytes of DRAM. Another 16 mega bytes of memory are added for error correction. Eight mega bit DRAM modules are used to construct the DBM unit. The upper left area contains the error correcting units for the DBM unit. The error correction units are needed because the total capacity of the DBM unit of the SIGHT-2 prototype system (16 PEs) is one giga bytes. The lower middle area is the WCS of the TARAI unit, and its size is 64 kilo words, each is 104 bits wide. The lower right area is the WCS of the MP unit. Its size is 64 kilo words each 96 bits wide. In the upper right area are the MP sequencer and pipeline registers.

Figure 15
Upper Board

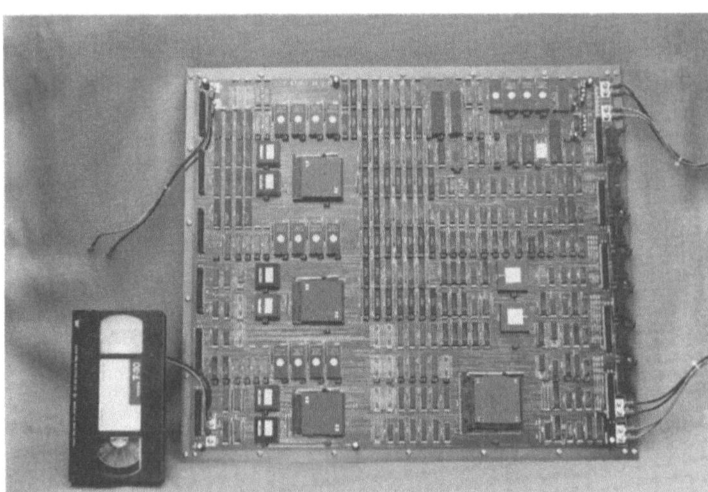

Figure 16
Lower Board

The lower board is shown in Figure 16. The major part of this board is occupied by the ALUs. The chip groups on the left of the board are made up of the floating point multiplier; WTL1164GCB[2], the floating point ALU; WTL1165GCB[2], EPROMs for the function table, and the TARAI controller VLSI. These units construct the TARAI unit and execute three dimensional vector operations in parallel.

The lower right area holds the MP unit. The connectors for the ports of the high-speed serial communication interfaces are located across the right edge of the board. The upper center area is the register files. Each register file is 256 kilo bytes.

The μP is located in the upper right. The monitor, the debugger and the initial microprogram for the MP unit are stored in EPROMs, the total capacity of which is 256 kilo bytes.

5. System Performance

The image shown in Figure 17 was used for comparing the performance of SIGHT-2 and a VAX11/780 with a floating point accelerator. The size of the image is 512 x 480 pixels. The surfaces of all spheres are reflective.

The assignment of image generation areas to each PE is shown in Figure 18, that is, vertical stripes the full height of the screen are assigned to each PE. The load balance of each PE is statically controlled. Static load balancing works well with increasing image complexity.

Measured system performance is shown in Figure 19. One PE is 10 times faster than the VAX, and 16 PEs are 158.8 times faster.

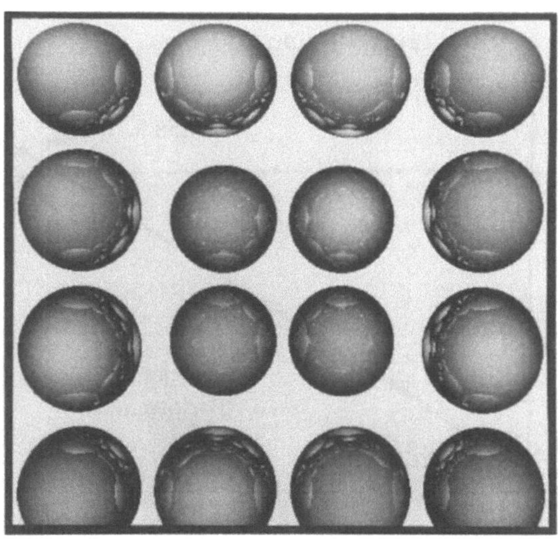

Figure 17 A Synthesized Image for Performance Comparison

---

2 Weitek Corporation

We define PE efficiency by the mean ratio of one PE execution time to the system execution time. Thus, PE efficiency is 99.2% in the ray tracing experiment. The overhead to wake up PEs and the load unbalance among PEs do lower the efficiency; however, from Figure 19, it is shown that this effect is small.

These results show clearly that the SIGHT architecture is well suited to efficiently execute the ray tracing algorithm in parallel.

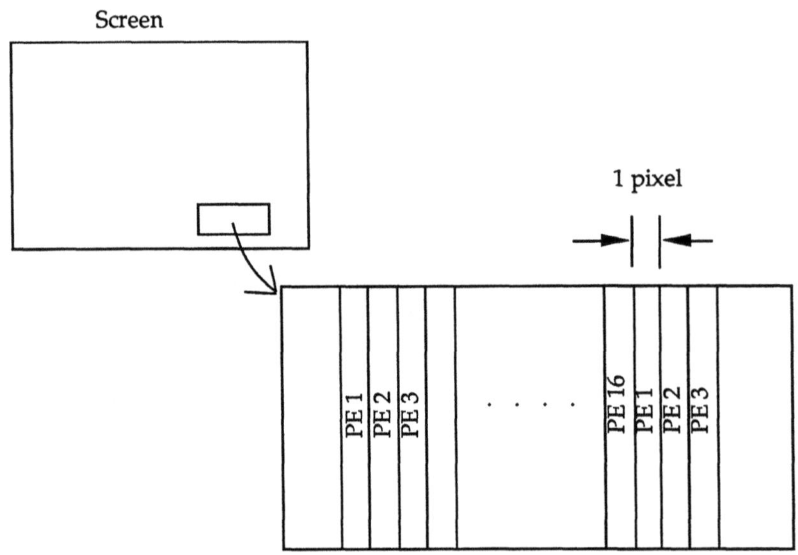

Figure 18 PE assignment

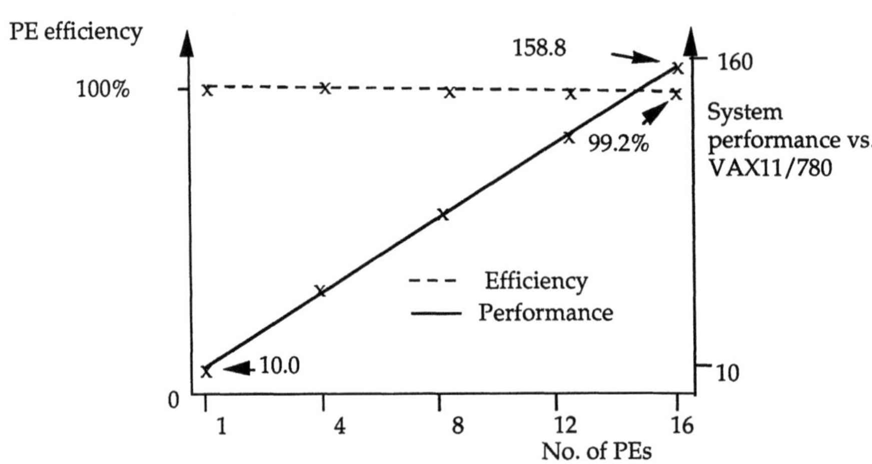

Figure 19 System Performance

6. Conclusion

The SIGHT is a multiprocessor system for the fast execution of ray tracing. To achieve high efficiency, three kinds of parallel architecture are introduced.

(1) Pixel level parallel processing,

(2) Operation level parallel processing,

(3) Functionally distributed parallel processing.

We designed two kinds of custom VLSIs; the TARAI Controller and Master Controller. With them, one PE can be constructed on just two printed circuit boards.

Each PE has a peak performance of 4.17 MFLOPS / 8.33 MIPS, and executes the standard ray tracing algorithm 10 times faster than a VAX11/780 with a floating point accelerator. The SIGHT-2 has 16 PEs and its overall performance is about 160 times faster than the VAX. Its peak performance is 66.72 MFLOPS / 133.28 MIPS. PE efficiency is 99.2% when executing the ray tracing algorithm.

Though SIGHT-2 has achieved significantly faster ray tracing, even higher performance is possible through hardware optimization with the same architecture.

Acknowledgements

We are grateful to Mr Hiroshi Kaneko, Mr Kei Takikawa and other colleagues in our section for their invaluable advice and encouragement.

Reference

1. Whitted, T.: An improved illumination model for shaded display. Comm. ACM 23(6), pp.343-349 (June 1980)
2. Nishimura, H., Ohno, H., Kawata, T., Shirakawa, I., and Ohmura, K.: LINKS-1 : a parallel pipelined multimicrocomputer system for image creation, Proc. 10th Annu. Int'l Symp. on Computer Arch., pp. 387-394 (1984)
3. Naruse, T., Yoshida, M., Takahashi, T., and Naito, S.: SIGHT - A Dedicated Computer Graphics Machine, Computer Graphics Forum (Journal of the European Association for Computer Graphics), Vol. 6, No. 4, pp. 327-334 (1987-12)
4. Takahashi, T., Yoshida, M., and Naruse, T.: Architecture and Performance Evaluation of the Dedicated Graphics Computer : SIGHT, Proc. of IEEE MONTECH '87 (Compint '87), pp. 153-160 (1987-11)

Viewing and Rendering Processor for a Volume Visualization System

A. Kaufman, R. Bakalash and D. Cohen

Department of Computer Science
State University of New York
Stony Brook, NY 11794-4400, USA

Abstract

The architecture and the hardware realization of the *3D Viewing and Rendering Processor* is presented. This processor is a component of the Cube architecture, developed primarily for volume visualization. The processor generates 2D shaded orthographic, parallel, and perspective projections of the volumetric image of n^3 voxels in $O(n^2 \log n)$ time. This performance is attributed to a unique skewed memory organization, a special ray projection bus, an extended viewing architecture, and a new ***congradient*** shading technique. A reduced-resolution prototype has been realized in hardware using printed circuit board technology and has been running in true real time. Currently, a VLSI version of the prototype is being tested.

1. Introduction

The *Cube Architecture* [8] is a versatile voxel-based architecture for three-dimensional (3D) volumetric graphics. The 3D objects are stored in a large 3D *Cubic Frame Buffer (CFB)* of voxels. A voxel, which is a unit volume cell, has a numerical value that characterizes its density, color, texture, opacity ratio, and the like. Cube is a multiprocessor system with several processors accessing the CFB to input, manipulate, view, and render the CFB images. The *3D Frame-Buffer Processor* acts as a channel for inputting 3D scanned voxel images, which are the primary sources of CFB data. Once the images are in the CFB they can be manipulated and transformed by the processor, which operates as a *voxblt* engine [7] (i.e., an extended 3D *bitblt* engine). It also acts as a monitor for 3D interaction. A geometric model is another source of CFB data. The *3D Geometry Processor* scan converts (voxelizes) 3D geometric models into their 3D voxel representation within the CFB, possibly intermixed with the scanned data. The *3D Viewing and Rendering Processor (VP3)* generates 2D shaded orthographic, parallel, and perspective projections of the CFB image of n^3 voxels in a conventional 2D frame buffer in $O(n^2 \log n)$ time. This paper focuses on the VP3 and its hardware realization.

In order to manage the huge quantity of voxels and still perform in real time, the VP3 is assisted by parallelism and speed-up mechanisms. Particularly, two unique features have been incorporated within the architecture: a skewed parallel memory organization, which permits the retrieval and storage of voxel rays in *arbitrary* directions, and a ray projection process, which exploits a multiple-write bus to select the voxel closest to the observer along the ray. These two key constructs of the Cube architecture reduce the 3D problem involving voxels to a 2D problem involving rays of voxels. Other key constructs of the architecture include three 2D buffers to assist in generating arbitrary projections and a shading unit that is gradient-based and table-driven.

2. Skewing Memory Schemes

A CFB of n^3 voxels is divided into n memory banks, each with n^2 voxels, in such a way as to allow, in one memory cycle, simultaneous conflict-free access to a full ray of n voxels from the CFB, accessing only one voxel from each bank. The practicality of the mapping from discrete 3D space onto the CFB memory banks depends primarily upon whether the mapping can be described by a simple formula. Consider first the case of beams, that is, rays parallel to a primary axis, as in orthographic projections. A voxel with discrete space coordinates (x, y, z) is mapped onto the k th bank $(0 \leq x, y, z, k < n)$ by the following simple skewing scheme

$$k = (x + y + z) \bmod n. \tag{1}$$

Since two coordinates are always constant along any beam parallel to a primary axis, the third coordinate guarantees that only one voxel from the beam resides in any one of the banks. The internal mapping (i, j) within the bank is simply

$$i = x, \qquad j = y. \tag{2}$$

With this mapping, conflict-free access to beams of voxels along any one of the six orthographic directions is provided. These six directions are along the positive and negative directions of rows, columns, and axles, that is, parallel to the primary axes $\pm x$, $\pm y$, and $\pm z$. A whole beam of voxels is retrieved from the skewed storage space by a process of de-mapping the voxel beam back into 3D discrete space. A voxel within bank k is de-mapped onto the voxel beam (e.g., $y = y_0$, $z = z_0$, i.e., parallel to $+x$ axis) in 3D discrete space in position

$$D_x = [k - (y + z)] \bmod n \tag{3}$$

along the beam.

When scanning the CFB beam after beam for viewing, the internal order of the banks along the beams of voxels changes. Actually, the bank index, which is the distance of a voxel along a beam from the viewing position, is either incremented or decremented by 1 modulo n when moving to the next beam. This change is controlled locally by the local address unit of the bank. The initial assignment of the bank indices depends only upon the viewing direction. Consequently, even the simple arithmetic involved in Equation 1 is avoided during the time-consuming viewing process.

Consider now the following linear skewing scheme of the CFB

$$k = (ax + by + cz) \bmod n' \qquad 0 \leq k, x, y, z < n' - 1 \tag{4}$$

where $n' \approx n$, and a, b, c, and n' are selected in such a way as to guarantee conflict-free access to all the desired directions. Equation 4 allows for rays in nonorthographic directions to be retrieved from the CFB conflict free. For example, the linear skewing scheme

$$k = (5x + 2y + z) \bmod 517 \qquad 0 \leq k, x, y, z < 517 \tag{5}$$

provides conflict-free access to 26 parallel directions, with projection time complexity identical to that of the orthographic projections, i.e., $O(n^2 \log n)$. The 26 projections are the six orthographic projections (along rows, columns, and axles), eight parallel projections along the major (principal) diagonals and antidiagonals, called *corner projections*, and 12 parallel projections along the minor diagonals and antidiagonals, called *edge projections*.

3. Projections

The ray projection mechanism of the VP3 exploits a special common bus, called the *Voxel Multiple-Write Bus (VMWB)* [8]. The projection logic is composed of n identical processing units. A full beam of n voxels fetched from the CFB along the viewing direction is placed into the processing logic, with each voxel in its associated processing unit, for the selection of the nontransparent voxel closest to the assumed observer. All the units holding nontransparent voxels "race" for the bus by comparing their depth index bit after bit, starting from the most significant bit, and the unit holding the voxel closest to the observer succeeds after $\log n$ steps. For every beam, the projection mechanism outputs two values associated with the closest voxel: its value (e.g., color) and its depth (e.g., distance). These two values are passed directly to the shading unit.

The memory, addressing system, and projection units have an overall modular structure comprising a sequence of identical modules indexed $k = 0, 1, \cdots, n-1$. These modules are interconnected with a barrel shifter allowing fast memory-to-memory transfer of beams in $O(\log n)$ time. Each module also includes a transparency unit, which is provided with transparency control parameters broadcast to all modules. Interval and depth control parameters are also broadcast and are compared with the locally computed depth measures (Equation 3) to generate depth sections or slices of the scene.

Arbitrary parallel projections can be generated in Cube in $O(n^2 \log n)$ time, by first transforming (rotating) the scene and then viewing it through a principal orthographic direction. With this technique, however, the CFB image is distorted each time a rotation is performed. Another technique that has been developed in Cube is ray casting [1, 6, 9-11]. It is suitable for both parallel and perspective viewing and has $O(n^3)$ complexity. Cube provides 26-, 18-, or 6-connected rays to accommodate objects in the scene that may have 6-, 18-, or 26-connected tunnels. Cube explores also the reconstruction of arbitrary parallel and perspective projections of convex objects from at least three orthographic projections and their respective depth buffers. The performance complexity is comparable to that of the orthographic projection, but may increase with the complexity of the scene. In addition, the CFB skewing storage schemes, discussed before, allow for rays in nonorthographic parallel projections to be retrieved conflict free. They can also be exploited by perspective projections, conceivably through a multiple, but limited, number of fetches per ray.

An extended viewing architecture, which uses three additional 2D buffers, has been designed to accommodate arbitrary parallel and perspective projections. Since there is no direct way to fetch arbitrary discrete rays from the CFB conflict free, instead, a whole projection ray plane is fetched first. The plane is fetched as n beams (parallel to a primary axis) in $O(n)$ cycles, and stored in a 2D temporary buffer. This buffer is then rearranged in 26-connected discrete ray representations that correspond to the projection rays, using a $O(\log n)$ complexity barrel shifter twice. Each projection ray is then fed into the VMWB for the selection of the first opaque voxel along that ray in $O(\log n)$ time. The total time for an arbitrary projection is $O(n^2 \log n)$, allowing real-time arbitrary projection. This is the same order of magnitude as that of the orthographic projection using the original Cube architecture.

4. Shading

The *congradient* shading technique [3] has been developed for the Cube architecture emphasizing real-time performance. It capitalizes on the advantages of both gradient shading [5] and contextual shading [2] by defining the surface orientation of a voxel as one of a finite set of local gradients and, consequently, can employ a fast table-driven mechanism. The quality of the shaded images generated by the congradient shading technique can be favorably compared to other conventional techniques. The location of the light source can be arbitrarily changed, which provides a kind of motion parallax.

Unlike other voxel-based systems [4, 6, 9, 11] that perform the shading in a separate post processor, the shading unit in Cube is an integral part of the projection process, and each projected pixel is pipelined through the memory access, the projection unit, and the shading unit. The Cube congradient shading consists of only two stages, and uses a single 2D array, e.g., a frame buffer. The projection stage generates an array of congradient values each of which holds the concatenated values of the color and the local gradient. The production time of a congradient value is interleaved with the projection time of the next pixel. Since the congradient calculation time for each pixel is much less than its projection time, no additional time is accrued for the shading. The next stage is the actual display which is done with a standard hardware where the congradient values are used as indices to the look-up table, which translates a congradient value to its shaded color based upon the light source parameters. As a result, this two-tier shading unit is faster than the alternative techniques, and is simpler for hardware realization.

Two congradient shading approaches have been implemented: unidirectional shading, in which the gradient is computed only from the depth of the horizontal neighbors, and bidirectional shading, in which both horizontal and vertical neighbors are employed. In both approaches the architecture is very simple and is suitable for hardware realization [3].

5. Implementations

A software simulation of Cube has been running successfully on color Sun-3, Sun-4, Silicon Graphics IRIS, and HP Turbo SRX workstations, using various resolutions, such as 128^3, 256^3, and 512^3. A reduced-resolution hardware prototype of 16^3 voxels, 8-bit each, has been realized in hardware and has been integrated under an IBM-AT with a VGA graphics display controller. The hardware realization consists of 16 modules, each of which is implemented as a custom-built printed-circuit board containing a CFB module with its mapping and de-mapping addressing mechanisms, its projection processing unit, VMWB unit, translucency control, depth sectioning, etc. Each printed-circuit board includes a module of CFB memory of size 256 (16^2) bytes. There are 24 ICs on each board; all of them are CMOS components. The interface board, which is built out of 40 TTL ICs, controls the whole assembly of modules using the common bus.

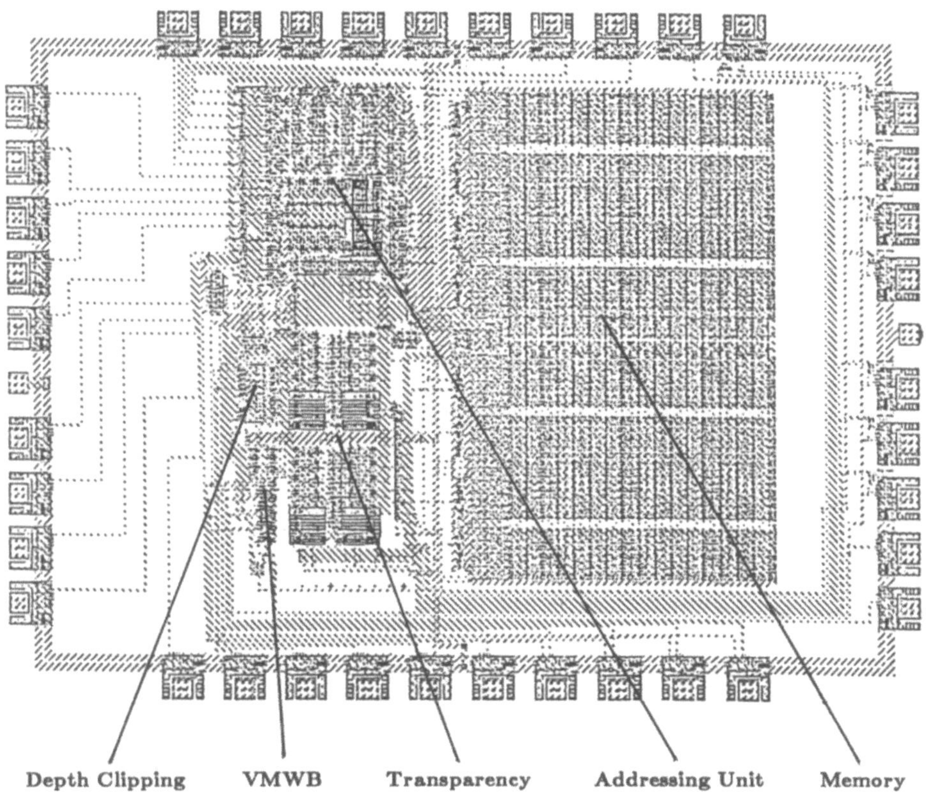

Figure 1: Layout of the VLSI design of the Cube chip

The prototype has been operating successfully in true real time. Measured performance figures on the prototype are 290 μsec for an arbitrary 3D rotation, 50 μsec for an orthographic projection, 330 μsec for an arbitrary parallel projection using rotation followed by orthographic viewing, and 160 μsec for arbitrary parallel or perspective projection using the extended viewing architecture. Estimated performance times for a 512^3 CFB resolution, using the same printed-circuit technology, are 511 msec for rotation, 62 msec for orthographic projection, and up to 573 msec for arbitrary projection. Using the VLSI technology will enhance the Cube speeds even further.

The modular structure of the Cube architecture is well suited for VLSI implementation. Currently a VLSI implementation is being pursued in which each board is fabricated as a single chip. Figure 1 depicts the layout of the VLSI chip design. The memory occupies the majority of the silicon area. The local addressing unit is the mapping and de-mapping circuitry of the skewed memory. The clipping unit is responsible for hither and yon clipping and slicing along the viewing direction. The transparency unit allows only those voxel values defined as opaque to be considered for the current operation. The VMWB is the unit that competes over the output bus for the selection of the nontransparent voxel closest to the observer. The various units of the Cube chip can be programmed by means of parameters which are broadcast by the central control unit.

The VLSI Cube chip has been designed on a Magic System, using SUN workstations, and fabricated with the MOSIS prototyping service. The Cube chip employs 2.0μ double metal CMOS/Bulk technology in a 40 DIP package. Figure 2 is a picture of the fabricated Cube chip. The whole IC assembly of 16 Cube chips occupies only part of a single board, which is shared with the interface unit.

The next prototype, which is under development, is a large real-time prototype for 256^3 resolution. The memory module will be separated from the logic and will be implemented using off-the-shelf memory chips. In order to reduce the fabrication cost and foster compaction of the hardware, two or more of the logic modules will be grouped together on one custom-built VLSI chip. This prototype will consist of 256 units of memory and logic chip each, assembled on several boards. The overall size will be about the same as the first 16^3 printed circuit prototype.

Acknowledgments:

We are grateful to S. Sherman and Z. Xu for their contributions to the design and building of the Cube prototypes. This work was supported by the National Science Foundation under grant MIP 88-05130.

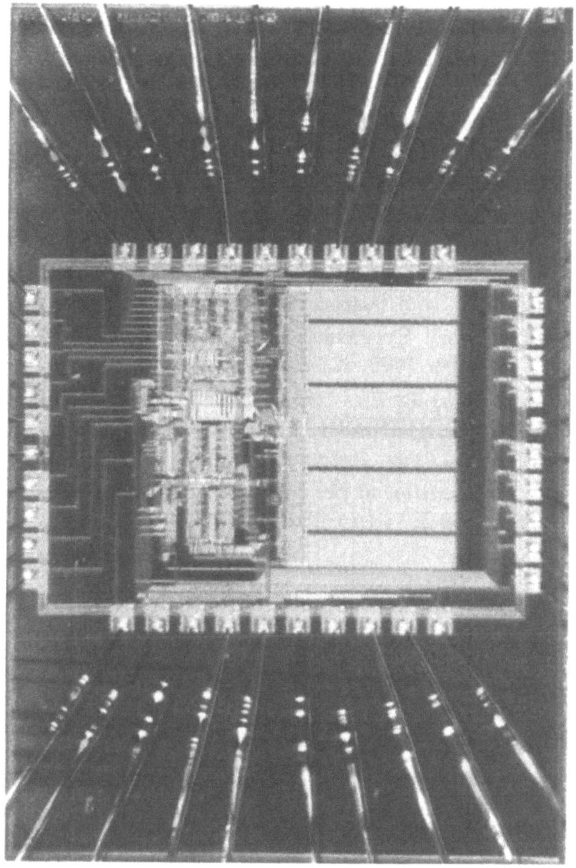

Figure 2: The fabricated Cube chip

6. References

1. Atherton, P. R., "A Method of Interactive Visualization of CAD Surface Models on a Color Video Display", *Computer Graphics*, **15**, 3 (August 1981), 279-287.

2. Chen, L. S., Herman, G. T., Reynolds, R. A. and Udupa, J. K., "Surface Shading in the Cuberille Environment", *IEEE Computer Graphics & Applications*, **5**, 12 (December 1985), 33-43.

3. Cohen, D., Kaufman, A., Bakalash, R. and Bergman, S., "Real-Time Discrete Shading", *The Visual Computer*, **6**, 1 (February 1990), 16-27.

4. Goldwasser, S. M., "A Generalized Object Display Processor Architecture", *IEEE Computer Graphics & Applications*, **4**, 10 (October 1984), 43-55.

5. Gordon, D. and Reynolds, R. A., "Image Space Shading of 3-Dimensional Objects", *Computer Vision, Graphics and Image Processing*, **29**, (1985), 361-376.

6. Jackel, D. and Strasser, W., "Reconstructing Solids from Tomographic Scans - The PARCUM II System", in *Advances in Computer Graphics Hardware II*, A. A. M. Kuijk and W. Strasser, (eds.), Springer-Verlag, Berlin, 1988, 209-227.

7. Kaufman, A., "The *voxblt* Engine: A Voxel Frame Buffer Processor", in *Advances in Graphics Hardware III*, A. A. M. Kuijk and W. Strasser, (eds.), Springer-Verlag, Berlin, 1989.

8. Kaufman, A. and Bakalash, R., "Memory and Processing Architecture for 3-D Voxel-Based Imagery", *IEEE Computer Graphics & Applications*, **8**, 6 (November 1988), 10-23.

9. Ohashi, T., Uchiki, T. and Tokoro, M., "A Three-Dimensional Shaded Display Method for Voxel-Based Representation", *Proceedings EUROGRAPHICS '85*, Nice, France, September 1985, 221-232.

10. Tuy, H. K. and Tuy, L. T., "Direct 2-D Display of 3-D Objects", *IEEE Computer Graphics & Applications*, **4**, 10 (November 1984), 29-33.

11. Uchiki, T. and Tokoro, M., "SCOPE: Solid and Colored Object Projection Environment", *Transaction of the Institute of Electronics and Communication Engineers of Japan*, **68-D**, 4 (April 1985), 741-748.

Presentation of the Cubi9000:
A Graphics System based on Inmos T800 Transputers

France Glemot

Caption[1], Rennes, France

1. Introduction

The Cubi9000 family includes a range of products from the 3D graphics terminal up to the 3D graphics workstation. The Cubi9000 when configured as a 3D graphics terminal connects to a host computer via a parallel interface from Digital Equipment Corporation. The Cubi9000 configured as workstation includes a UNIX[2] based mini-computer with a 32 bit microprocessor, mass memory, device handlers (mouse, tablets, encoders) and communication drivers to external systems.

2. General hardware overview

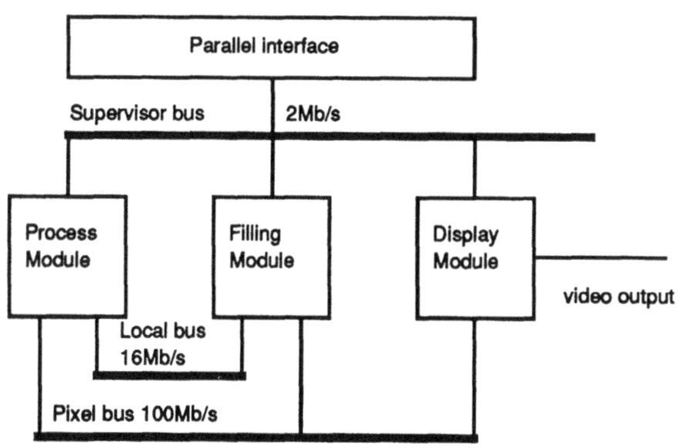

The basic hardware comprises three modules communicating via three different buses.

---

[1]The company Caption is a subsidiary of the company Telmat which manufactures telecommunication systems and minicomputers. Caption specialises in 3D image synthesis, it develops, builds and sells 3D graphics systems.

[2]Unix is a trademark of ATT Bell Laboratories

2.1 Display module

The Display module includes a frame buffer and a video interface:

The frame buffer comprises 40 1024*1024 bit-planes with a 16 bit hardwired Z-buffer providing 16 million simultaneous colours and uses a conditional writing-by-depth test for hidden part removal. The image formats may be dynamically programmable from 512*512 to 1024*1024 pixels.

The video interface provides an interlaced or non-interlaced video output, and an external synchronisation (genlock) is available for the video formats. It is possible to mix two plane images in depth at the video pixel rate, before the input to the digital to analog converters. Furthermore one of these planes can receive video images via a digitizer entry with blue box capabilities. So real and synthetic images can be mixed in real time.

2. 2 Filling module

The filling module comprises one or two polygon generators performing polygon shading (Gouraud shading at 20 Mpixels/s per generator). Each polygon generator consists of 2 boards, the first one computes the interpolation of colour and depth along the edges and the second computes the interpolation of colour and depth of the horizontal lines filling the polygons. These two boards are completely hardwired and offer 120 Mips.

2.3 Process Module

The process module comprises one or two Display Processor Units (DPU). The DPU is a parallel work flow board with several INMOS transputers, the number of on-board processors depends upon the processing power required. This module contains the display list and handles the image generation algorithms.

3. Architecture of the Display Processor Unit

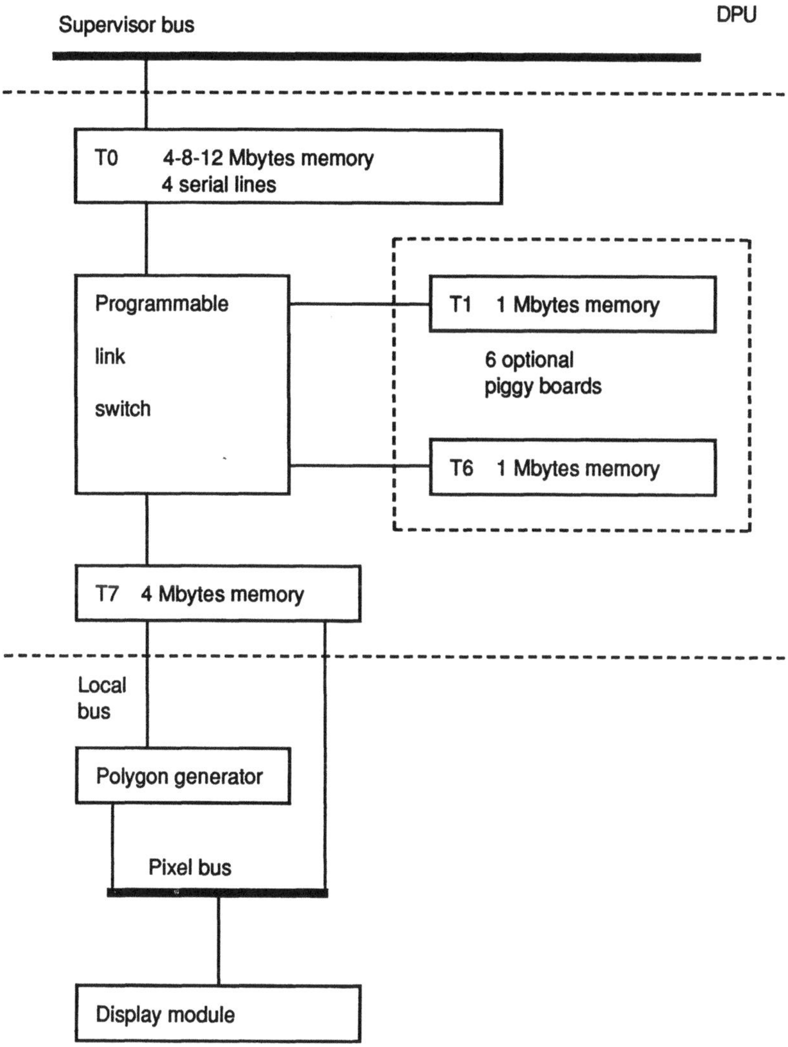

Block diagram of the Display Processor Unit associated with the Polygon Generator and the Display Module

3.1 Description of the transputer T0

The transputer T0 is non-optional, its hardware resources are the following:

- 4, 8 or 12 Mbytes dynamic RAM memory accessible by the Host via the Supervisor bus. This bus is shared between the Host and T0
- access to all other boards, via the Supervisor bus, in order to configure the Cubi9000
- four RS232 serial lines for connecting graphics peripherals
- two synchronization signals: an interrupt signal to the host and a video top-of-frame signal from the video board of the display module
- four links: one dedicated to the host (for program loading and urgent messages), the other three being connected to the programmable link switch (for communication with the other transputers).

The main tasks of T0 are: decoding of the received messages from the host, execution of the corresponding function, management of the graphics peripherals, sending back information to the host, processing geometric transformations (for Gouraud shading) and, if the DPU has only 2 transputers T0 and T7, it is responsible for high quality rendering effects (Phong shading, texture mapping, solid textures...)

3.2 Description of the transputer T7

The transputer T7 is non-optional, its hardware resources are the following:

- 4 Mbytes dynamic RAM memory (not shared and not accessible from the Supervisor bus, the Local bus or the Pixel bus)
- access to the Local bus in order to send data to the polygon generators
- access to the Pixel bus in order to write/read pixels directly in/from the frame buffer of the display module
- four links connected to the programmable link switch, for communication with the other transputers.

The tasks of T7 are mainly to collect data processed by the other transputers in order to send them to the polygon generator (for Gouraud shading) or to the frame buffer (for Phong shading). It performs anti-aliasing post-processing, using a choice of two techniques: oversampling and sub-precision for pixels belonging to the edges. It computes the features of polygons projected on the screen space (for Gouraud shading) and, if the DPU has only two transputers (T0 and T7), it produces high quality rendering effects (Phong shading, texture mapping, solid textures...)

3.3 Description of the transputers T1-T6

The transputers T1 to T6 are optional. They are equipped with 1 Mbytes dynamic RAM memory which is not shared and cannot be accessed by other processors. They have 4 links connected to the programmable link switch for communication with the other transputers.

The tasks of T1 to T6 are mainly: processing geometric transformations, computing the features of polygons projected on the screen space, processing high quality rendering effects (Phong shading, texture mapping, solid textures...).

The six optional piggy boards allow the processing power to be increased from 20 Mips/3 Mflops to 80 Mips/12 Mflops per DPU.

The reset, analyse and error signals of each transputer in the network are linked together by a daisy chain mechanism.

3.4 Description of the programmable link switch

This programmable link switch allows any link from one transputer to be connected to any link of any other one (link0 from T0 is not available for this purpose). The 8 transputers are linked together via this programmable link switch, allowing an optimum configuration to be provided for each application. Thus the Cubi9000 is actually a multi-mode device, able to support efficiently several image synthesis computing techniques (from Z-buffer to ray tracing and from polygon to octree modeling).

4. Software functionalities

The graphics library is written in the C language, but the calling level may support other languages such as Fortran, Pascal, ADA.

The graphics library comprises four independent parts: initialisation and configuration of the Cubi9000, management of the display list, management of the graphics peripherals and 2D functions.

4.1 Initialisation and configuration of the Cubi9000

The functionalities are the following: open/close a session without modification to the configuration or with default values, load colour look-up tables, select the video format, parameters of the display monitor, colour and depth of the background, erase/display a plan of the frame buffer, select the visualization mode (real time or not, mixing of 2 planes of the frame buffer or not), digitise a static image (colour slide, photography...) or dynamic images (camera, magnetoscope...) for the whole of an image or just those parts of it detected by a blue box.

4.2 Management of the display list

The display list comprises of 5 components: the computing features of the scene, the cameras, the objects, the light sources, the data tables (reflectance, texture...).

An entire component of the display list may be: added, deleted, duplicated, or substituted. A field of a component may be set or get.

4.3 Computing features of the scene

These computing features are the following:

- computing mode (normal, designation, picking)
- aliasing mode (none, oversampling, sub-precision)
- geometric description level (bounding box, object)
- rendering level (wire frame, Lambert, Gouraud or Phong shading, specific effects).

Cameras

The cameras are defined by:

- projection (orthogonal, conic_z, conic_1/z, conic_a+b/z)
- position of the observer
- features of the viewport (window on the screen)
- features of the window of projection
- features of the mask on the window of projection.

Objects

The objects are defined by:

- global attributes:
    - drawing flag (define whether or not the object must be taken into account)
    - bounding box features
    - position and coefficient of reduction/expansion
    - colour flags and colour coefficients
    - smoothing flags and smoothing coefficients (colour, geometric and arbitrary smoothing)
    - aspect flags and aspect coefficients (coefficient for Phong shading: - diffuse, specular, brilliance, transparency: min, max, thickness, and reflectance)
    - flags for the geometric description and rendering levels
    - flags for the processing of backfaced polygons (normal, interior, or sections which give different views of an object modified by a front clipping plane, according to a normal inside-outside description or outside only description
- polygon description table
- vertices table

- polygon colour table
- aspect table.

Light sources

An ambient light source will always exist. The light sources are defined by type (sun, point-source, spot), the intensity, the position (suns are located at infinity), the direction (not relevant for the point-source), the colour, and the cone angle (relevant for spot only).

Data tables

Two tables may be used: the reflectance table and the texture table.

4.3 Management of the graphics peripherals

Four RS232 lines on the DPU allow it to be connected to: an alphanumeric terminal, a 2D tablet, an 8 encoders board, a magnetoscope (format BVU, BETACAM, 1 inch), and to any other peripheral programmable via an RS232 line.

The reading modes are compatible with those of Phigs: Sample, Request, Event. The data issued from a graphics peripheral (other than a magnetoscope) may be sent directly to the host, processed on the DPU and sent to the host, or processed and used in order to modify a field of a component of the display list between two image computations. This is done by the DPU, the host has just to send the command for image computations.

The alphanumeric terminal may be initialized, activated/stopped, read and written. The 2D tablet may be initialized, activated/stopped, configured (operating mode, resolution, sample speed, origin, size..) and read. The 8 encoders may be initialized, activated/stopped, configured (sample speed, minimum threshold) and read. The magnetoscope may be initialized or instructed (play, stop, forward, rewind, read/write/go_to a time code, record, standby). An interface between the magnetoscope and the Cubi9000 based on PC computer is necessary.

4.4 2D functions

All the components of a pixel R, G, B, and Z or only one of them may be read, written or written conditionally upon its Z value, in a rectangular space of the frame buffer. Cursors and reticles may be defined, selected, positioned, moved, erased. Coloured vectors and polygons may be drawn in any plane of the frame buffer. A character font may be defined, loaded, selected. By default a standard font is defined, loaded and selected. A string may be written in any plane of the frame buffer using the following features: colour, position, direction, thickness and reduction/expansion for character height, width, spacing.

5. Software architecture

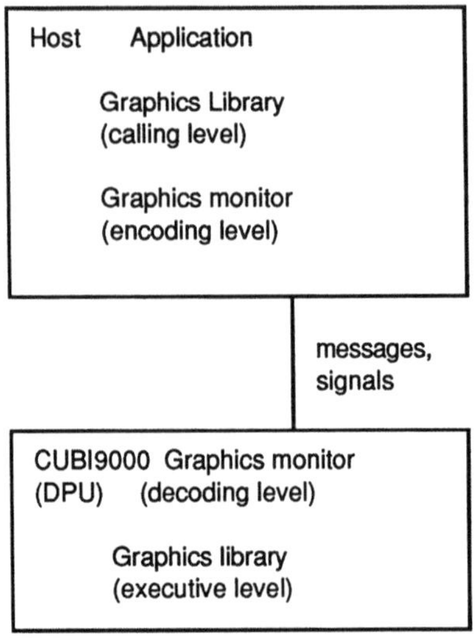

The communication between the host and the Cubi9000 is based on messages and synchronization signals. The application is located on the host for the obvious reason of portability, but some parts of it could be exported in order to complement the library, even for non graphics processing. The calling level of the graphics library must be located within the application, it performs some coherence checking on the parameters. The executive level must be efficient, and for this reason it is implemented in firmware.

The graphics monitor has to transfer the graphics commands from the host to the DPU, to synchronize them and to transfer data from the DPU to the host.

The urgent messages are sent to the address of link0 of T0, they are scanned automatically and processed by a concurrent process. The mechanism of urgency is used when an error occurs or when the user wishes to terminate immediately an animation sequence. The mechanism of urgency stops the execution of the current command and re-initialize some variables which are shared between the host and the DPU.

Another concurrent process manages the graphics peripherals. The tasks of the encoding level of the graphics monitor are to collect commands and their associated parameters from the library and packet them up in a format understood by the DPU. It is required to wait if the command input fifo on the DPU is full and otherwise send the packet to that fifo. If a reply is required, it waits for the end of the execution and, in which case, gets data from an output data buffer on the DPU, it unpackets them into a format understood by the host and stores them in the place defined by parameters.

The tasks of the decoding level of the graphics monitor are to wait if the command input fifo is empty, decode the packet, performing the corresponding function or using the executive level of the graphics library to send it to the underlying network. Finally, if a response is required, the data is put into the output data buffer and then a synchronisation signal is sent to indicate to the host that the execution of the command is finished.

The command input fifo and data output buffer are shared, at the graphics monitor level, by both the host and the DPU (T0). Conflicting accesses are resolved mainly by the rule that the variables shared may be read by both, but must be written by only one. For instance, the host is not allowed to reset the status of the DPU itself, and has to send a urgent message to the DPU.

When a command does not need a synchronization signal at end of its execution, both the host and the Cubi9000 can work in parallel. Furthermore the command input fifo serves to regulate the flow of commands between the host and the Cubi9000.

6. Constructing load balanced systems

6.1 Some general rules

To achieve good performance, a program may be split into several processes implemented on a network of transputers. Communications within such a network use the links as channels.

6.1.1 Simple architecture

Consider first a case in which there are two transputers running processes P1 and P2, if one process is (much) faster than the other, one of the processors will be under utilised, but we will now assume the system is correctly load balanced. The compute times and communication periods are denoted, respectively, CPT and CMT.

Load balanced systems imply CPT1 = CPT2 (to a close approximation) and synchronized communication implies CMT1 = CMT2.

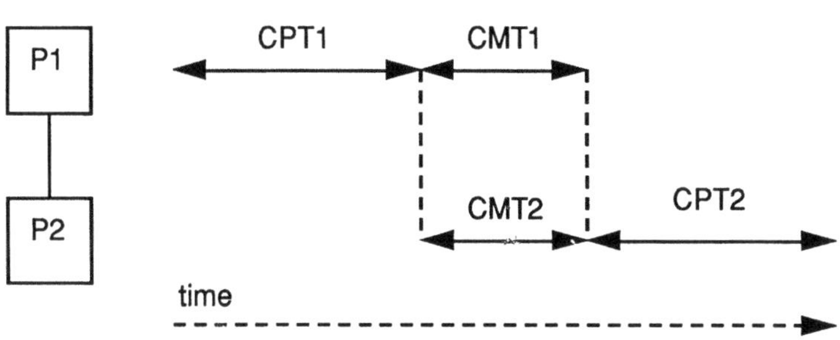

The data processing rate, to an external observer, is CPT1 + CMT1, for a non-balanced system we have Max(CPT1, CPT2) + CMT1.

6.1.2 An improved architecture

We now consider the same problem, but with two additional processes: a transmitter process T1 on the first transputer and a receiver process R2 on the second transputer. T1 and P1 operate in parallel, communicating via two buffers and using a flip-flop mechanism between the two buffers. A similar situation exists for R2 and P2 on the second transputer. We may consider that T1 and R2 introduce a small offset in the execution of P1 and P2 because each of them uses a DMA independent of the ALU, the offset is due to the fact that the internal bus must be shared between the DMA and the ALU and the fact that a time slicing is necessary between P1 and T1 (and between P2 and R2). The effect of time slicing is, however, very small. We denote these offsets as O1 and O2, and assume that any difference between O1 and O2 is small.

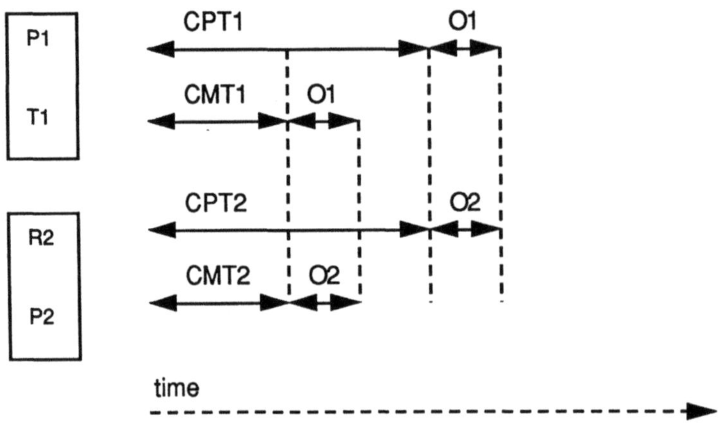

The perceived external data processing rate is CPT1 + O1, and for a non-balanced system we have Max(CPT1+O1, CPT2+O2).

The results are better than those obtained above and for a longer pipeline we can generalize in the following way:

> On the first transputer, the processes are P1 and T1 (and maybe an R1 for the external world); on the following transputers, the processes are Ri, Pi and Ti; and on the last transputer, the processes Rn and Pn (and maybe a Tn for the external world).

Ri and Ti are secondary processes which arise due to this manner of implementation, and only the Pi processes perform useful work for the application. This is the reason why the processes Pi need not wait for the communication processes Ti and Ri.

We have CPTi $\geq$ CMTi for Ti and the same for Ri.

6.1.3 Advantages of the pipeline architecture

Splitting the program into several processes results in each of them being shorter and thus they may be held within the internal memory of the transputer. Generally this also applies to the data. The internal memory accesses are much faster than external memory accesses (by a factor of 2 or 3 times). The dispatching of commands and data and the collection of results are simpler than within a parallel architecture.

6.1.4 Advantages of the parallel architecture

The parallel architecture does not suffer the time delays of a pipeline architecture. incurred when the pipeline is being loaded and emptied. The parallel architecture is more efficient than a pipeline architecture because it uses fewer processes , implying less time slicing and fewer data transfers. If splitting the code results a large amount of data to transfer then the rule that a computing process must wait for communication may not be hold.

6.1.5 Advantages of the mixed (pipeline and parallel) architecture

In practice, transputer based systems may have a mixed architecture in order to communicate easily with the external world and to achieve efficiency. Nevertheless with the same number of transputers several architectures are possible, as is shown below by considering three application examples.

6.2 Three examples of applications

6.2.1 Real time simulator

In this case the DPU performs the 3D to 2D transformations and determines the edge slopes for each polygon; the polygon generator being responsible for filling these polygons.

There are three steps in the basic algorithm: the visibility test (backface removal) and pre-clipping, the 3D to 2D transformation with clipping and illumination of each vertex, and the slope processing (in which the gradient values of X,Y,Z,R,G,B are determined). The results are communicated to the polygon generator.

The resulting transputer network architecture is organized in three pipelines according to the following two rules: the work load for T1 to T6 must be similar, and the work load for T0 and T7 must be less than 33% of the work load of others transputers.

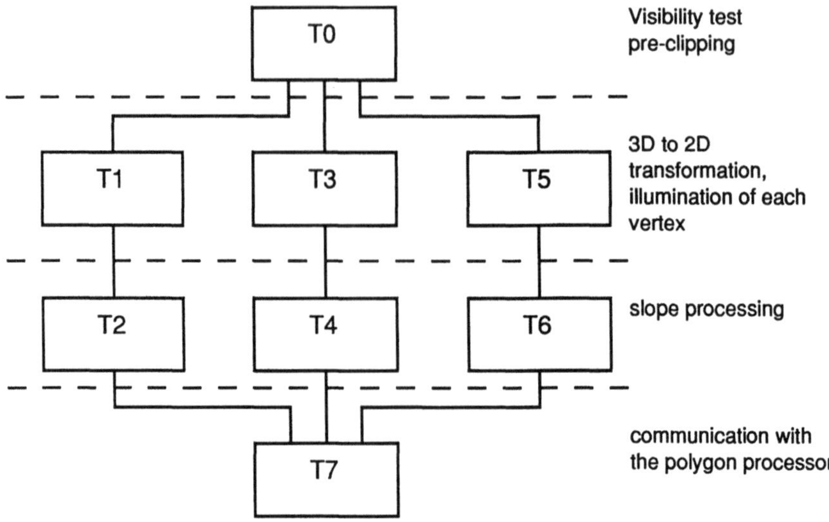

This logical architecture is similar to the physical architecture, each transputer having a receiver process R, a transmitter process T and one or several computation process(es) P.

The Cubi9000 can process and display from 4000 to 40000 3D Gouraud shaded polygons per second (with an average of 300 pixels/polygon).

6.2.2 High quality images

In this case the polygon generator cannot be used, and the DPU has to compute each pixel value, so the program is larger than that for Gouraud shading. Furthermore there are additional specific effects which require large data tables (reflectance, texture...), with the result that the internal memory of the transputers is too small to hold the program and data.

The amount of computing power required to compute illumination and shading at each pixel is much greater than the computing power needed for geometric transformation, with the consequence that the process cannot be split at this level. It is not possible to make a split between illumination and shading, without having to transfer a large amount of data, and incurring large communication overheads.

Furthermore in the architecture shown below, a single source program can be used with 2, 4, 6 and 8 transputers on the DPU.

The basic algorithm is decomposed in three steps: the visibility test and pre-clipping, the 3D to 2D transformations with clipping, illumination and shading of each pixel, and the communication with the display module.

Using the rules previously defined, the logical network architecture is:

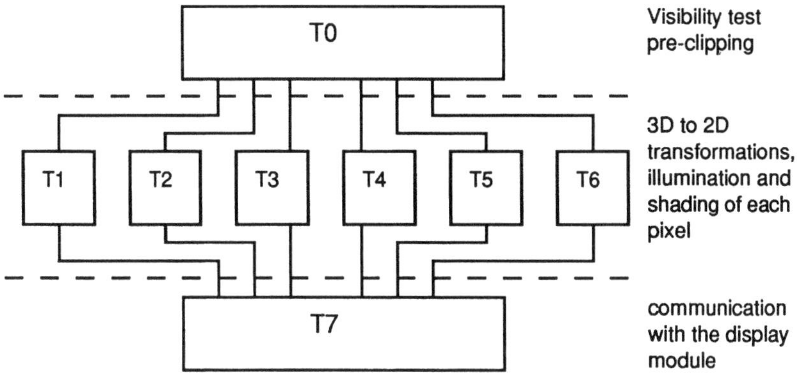

This diagram represents only a logical architecture, in practice T0 will have only 3 links connected to the programmable link switch. So there is a by-pass mechanism on T1, T2 and T3 in order to dispatch data to the computing process within the transputer or to the receiver process on the transputers T4, T5 and T6. On T4, T5 and T6 another by-pass is used to transmit processed pixels received from T1, T2 and T3 directly to T7 or to transmit the data to the computing process within the transputer. Each transputer has a receiver process R, a transmitter process T and one or several computation process(es) P.

The physical architecture is the following:

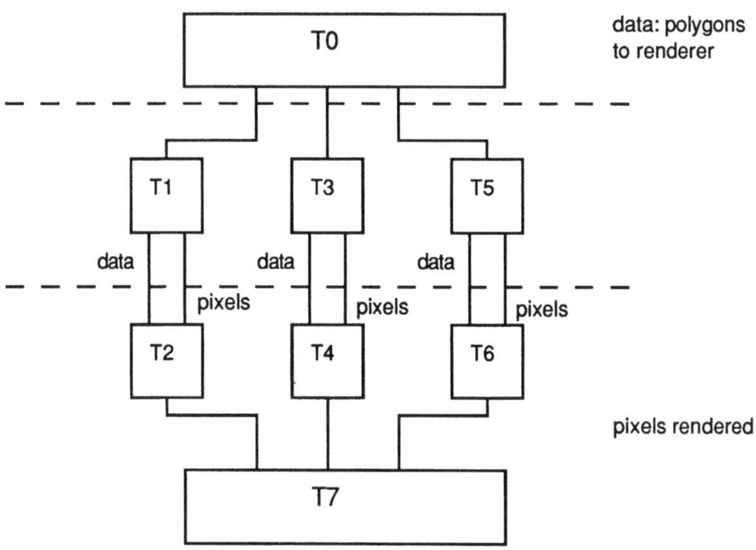

The Cubi9000 can process and display from 3500 to 40000 Phong shaded pixels.

6.2.3 Octree

For octree applications another development based on a further board is now considered. This board is based on two modules of 4 transputers each, and connected via 7 links to a DPU equipped with the two transputers T0 and T7 (note that link0 of T0 is not available for such a purpose).

The architecture chosen is based on a cube with 8 transputers, one at each vertex. The advantages of such an architecture is that it is similar to the octree organization which makes the data dispatching algorithm easier to implement, and there is a smaller number of connections between the transputers within the cube thereby reducing the transfer time for data.

It will be observed that most operations needed involve modifications to the octree dispatching because they alter the tree organization. Such operations are the creation of the tree, creation of objects (prosthesis...), segmentation (detection of structures corresponding to an organ, tissue...), rotation or translation of the tree (modification of the reference).

The only difference between the logical architecture and the physical architecture is the link between the transputer T0 on the DPU and the transputer T0 on the other board. There is a logical link from T0 on the DPU and T0 from the other board, but physically this link uses T1 (on the other board).

The logical architecture is the following:

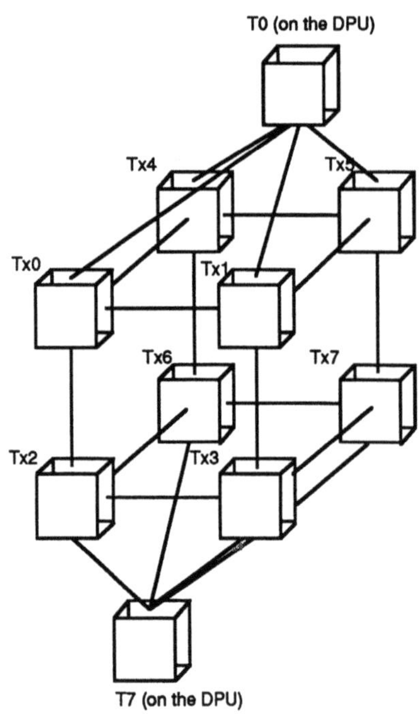

7. The transputer and Occam[3]

7.1 Introduction

This section presents an overview of the transputer family and a description of the main concepts in Occam with their associated implementation on transputers.

7.2 Overview of the transputer family

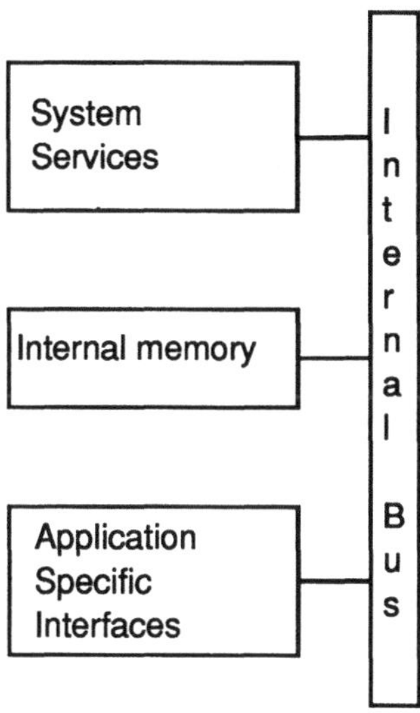

7.2.1 System Service

The main signals are the following:

- reset: this input signal resets the transputer. On the T414 and T800, this signal also resets the configuration of the external memory interface (code and data located on external memory are destroyed)

- analyse: this input signal resets the transputer prior to debugging. On the T414 and T800, the configuration of the external memory interface, code and data located on external dynamic RAM memory are not changed.

- error : this output signal is an indication to the external world when an error occurs at execution time. In the T414 and T800 an additional error

[3]Occam and INMOS are trademarks of INMOS

flag is internally set for debugging facilities. Error checking possibilities are, for instance, array boundary overflow, and arithmetic overflow.

- boot: this input signal permits booting from a external ROM memory. Another possibility is to boot via a link.

7.2.2 Internal Memory

The on-chip memory is 2 Kbytes in the T414 and 4 Kbytes in the T800. No caching facilities are provided. The accesses to code and data located in this memory are much faster than to those located in the external memory. The programmer may force certain code and data to reside in a specific address area, for instance in this internal memory.

7.2.3 Application Specific Interfaces

In order to allow the design of an interface with a minimum of external components, the transputers may include a controller for SCSI disk, graphics or memory. Timings and standard control signals are programmable by software.

7.2.4 Internal Bus

This bus multiplexes 32 bit addresses and 32 bit data, in both the T414 and the T800,

7.2.5 Processor

The T414 has an integral 32 bit integer Arithmetic and Logic Unit (ALU). The T800 has a 32 bit integer ALU and a 64 bit IEEE floating point ALU. The 30 MHz T800 offers 10 Mips and 1.5 Mflops.

7.2.6 Link Interfaces

A serial input line and a serial output line are associated with each link interface. The T414 and the T800 provide 4 link interfaces. A link is a point to point connection from a transputer family chip to a transputer family chip. Within a network, transputers are connected via links.

Point to point connection links have many advantages over multiprocessor buses:

- no contention for the communication mechanism,
- no capacitive load penalty as transputers are added to a system,
- no saturation of the communication bandwidth, and furthermore an higher, total communication bandwidth in the system.

These advantages are independent of the number of transputers in the system.

On the T414 and T800, the link interfaces incorporates two separate DMA devices, one for each direction. Thus communication and processing can take place independently, with only the internal bus being shared.

Link adaptors are part of the transputer family, they allow transputer links to be interfaced to a non-transputer sub-system. They convert data to and from a link to an 8 bit parallel port, taking into account the specific communication protocol of the links. Link adaptors may have two operating modes depending on whether they work like two uni-directional parallel ports or single bi-directional parallel port. The features and advantages of link adaptors are the following: data reception is asynchronous, which implies that communication is independent of the clock phase and transfers can take place in both directions simultaneously (but with some speed penalty).

7.2.7 Peripheral Interfacing

Three methods may be considered:

- a transputer with 1,2,3 or 4 peripheral control transputers
- a transputer with 2 link adaptors
- a transputer with peripheral addresses space mapped in the memory space.

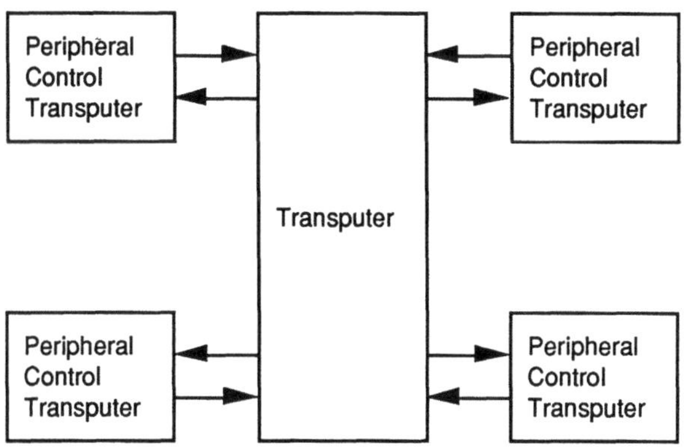

a transputer associated with 4 peripheral control transputers

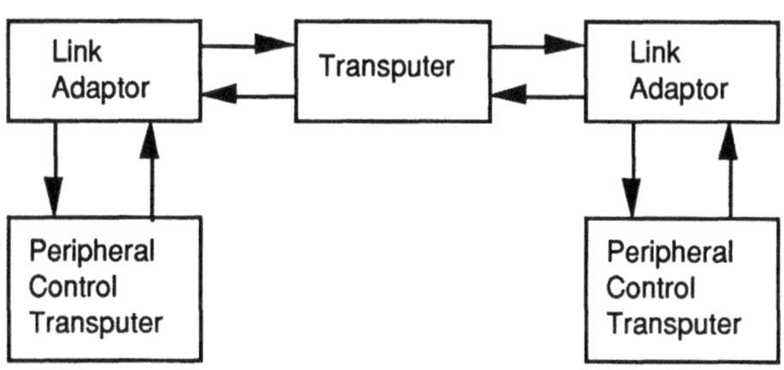

a transputer associated with 2 link adaptors

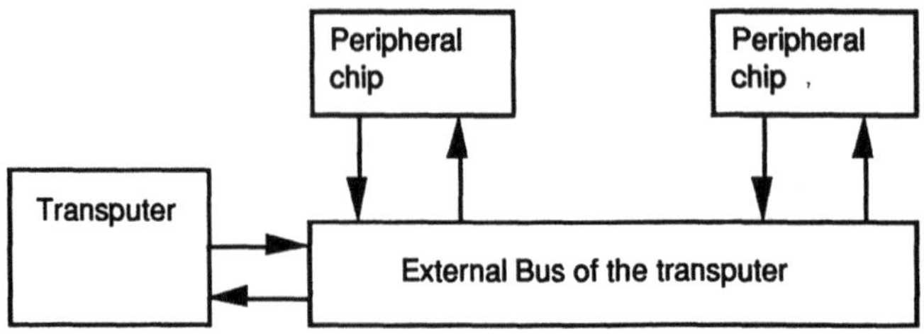

a transputer with peripheral addresses space mapped in the memory space (the peripheral is controlled by memory accesses issued as a result of a port inputs and port output, cf next paragraph Concepts in Occam)

7.3 Concepts in Occam

Occam is a programming language for the transputers, it is a high level language giving a maximum program efficiency and is able to exploit the special features of the transputers. Occam and the transputer were designed together. Occam can be used as an harness to link modules written in standard languages (C, Fortran, Pascal...). This harness includes all information concerning the configuration.

Occam provides a framework for designing concurrent systems using transputers.

7.3.1 Processes

The software building block in Occam is called process. A system is an interconnected set of processes communicating via point to point channels. A hierarchically structured system is a set of processes designed themselves as a set of (sub)processes. Each process is an independent unit of design, completely specified by the messages it sends and receives. A process starts, performs a number of actions and then terminates. An action may be a set of any number of sequential or parallel processes. Procedures and functions are processes.

When two processes of different priority are concurrent, the process with the larger priority must stop (because it is waiting for a communication for instance) or complete, before than the other may run.

Concurrent processes with low priority are periodically timesliced (generally the running period is some hundreds of microseconds).

7.3.2 Channels

Channels are the logical medium used by parallel processes in order to communicate values, no variables are shared between such parallel processes. For each communication a channel, a transmitter process and a receiver process must be defined; the communication occurs only when the both processes are ready.

When two processes communicating via a channel are implemented on the same transputer, only data moves within the memory are performed.

When the channel is implemented on a transputer link, a protocol is used during the communication, its features and advantages are the following:

- a byte acknowledge, which provides reliable communication between a slow and a fast transputer and word length independency
- an anticipated acknowledge, which implies a continuous data flow
- a standard communication frequency of 10 Mhz for all the transputer family (nevertheless other frequencies exist: 5 Mhz, 20 Mhz..)
- no clock phase dependency, which imply that communications may happen between two independently clocked systems.

Communication via channels is synchronized automatically, no additional explicit programming is necessary. Furthermore the structure of a program which communicates between two processes is independent of whether the processes are executed on the same transputer or on two different transputers.

Communication via any channel may occur concurrently with communication via other channels and with program execution.

If the program is written in a language other than Occam, then a run-time system is provided which provides input/output to Occam channels (and which conforms to the link protocol if links are used).

Peripheral access may use ports, which are extensions of the channel mechanism, for a block of information consecutively stored in memory. Nevertheless there is no synchronization mechanism associated with a port input and output. So a value read by a port input may depend upon the time at which the input was executed, and inputting at an invalid time would produce unusable data.

7.4 Implementation of the Occam on the transputer

The transputer and Occam were designed together. Hardware features are provided to support the special features of Occam. Whatever the number and kind of transputers used, the Occam level interface remains the standard interface.

A transputer link interface may be used to implement two channels, one for each direction. A program running on a transputer is called a process. All transputers incorporate a timer providing a clock which can be used by any number of concurrent processes for internal measurement or for real time scheduling. The hardware interface between a dedicated sub-system and a transputer based system has to use one or more links. During the development stage, the sub-system could easily be simulated by an Occam implementation of the associated process.

A system can be built with a single transputer independently of the number of processes, or with a network consisting of a large number of transputers without any topological restrictions, excepting that the T414 and T800 transputers have only four link interfaces. In both cases the identical source program can be used. So that an Occam program may run on a single transputer when the cost must be optimised or on a network of transputers for performance optimisation.

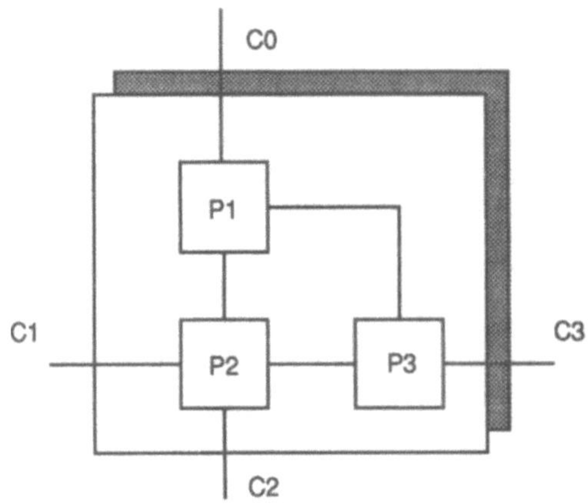

Optimisation for cost:

A program on a single transputer with 3 processes P1, P2 and P3, 4 links to the external world, and 3 internal channels between the processes (data moves within the memory)

The transputer shares its time between the 3 processes

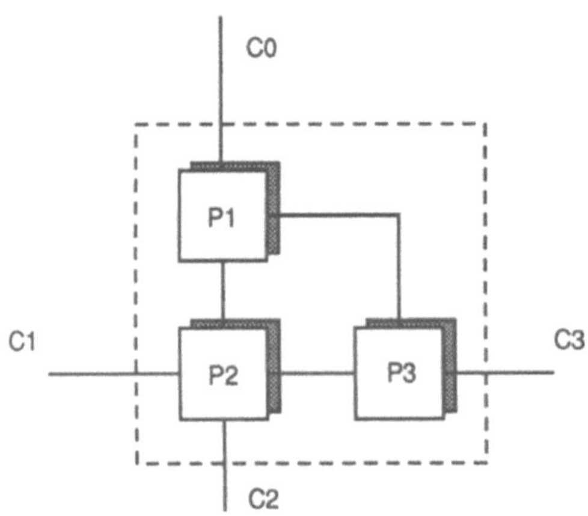

Optimisation for performance:

A program on 3 transputers with 1 transputer for each process, 4 links to the external world, and 3 links for communication between the transputers.

Each transputer executes a single process

A Virtual Memory System Organization for Bit-Mapped Graphics Displays

Anthony C. Barkans

Schlumberger Technologies, Inc.*
CAD/CAM Division Ann Arbor, Michigan USA

ABSTRACT

Described is a display sub-system, designed for support of a very high speed rendering engine. It provides high-performance graphics to an environment that consists of a hierarchy of resizable windows. The concept of virtual memory has been applied with the organization of the virtual to physical address spaces having a unique mapping that fits the organization of a bit-mapped graphics memory display.

CR Categories and Subject Descriptors: I.3.1 [Computer Graphics]: Hardware Architectures - raster display devices; I.3.3 [Computer Graphics]: Picture/Image Generation - display algorithms.

Introduction

VIEWS, is a new graphics memory system developed to support rendering over one million triangles per second in a full X Windows environment. The limitations of current graphics memory architectures in this environment will first be reviewed. These limitations necessitated a new approach based on the specific graphics memory requirements of high-performance rendering in X. The resulting VIEWS architecture is an innovative and practical solution which exceeded these requirements.

Current Graphics Memory System Design

The designs of hardware memory systems for bit-mapped computer graphics have often followed general-purpose CPU paradigms. The three methods developed for mapping a CPU's address on to its physical memory are direct mapping, segmented virtual memory, and paged virtual memory. All of these methods have more recently been applied to graphics systems as graphics capabilities increase.

Direct mapping of graphics memory to the rendering engine's address space is the textbook example [4] commonly used in current commercial systems. This approach has the lowest cost for a minimum system designed with a frame buffer containing only enough memory to refresh the CRT. Such systems have two performance problems when used in a high performance window environment:

---

* Currently with: Hewlett Packard, Graphics Technology Division, 3404 Harmony Road, Ft. Collins, Colorado USA

- The rendering engine must clip pixels to the boundaries of windows. These boundaries can be almost arbitrarily complex. As rendering speed increases and more windows are used simultaneously, clipping at full rendering speed becomes very costly.
- As the user moves and deletes windows, exposure events must be sent by the window manager to the graphics applications in other windows. These applications must then redraw at least portions of their windows, using both CPU time and graphics throughput.

Figure 1 shows a single memory plane of such a system and how its contents map to the display surface of the CRT.

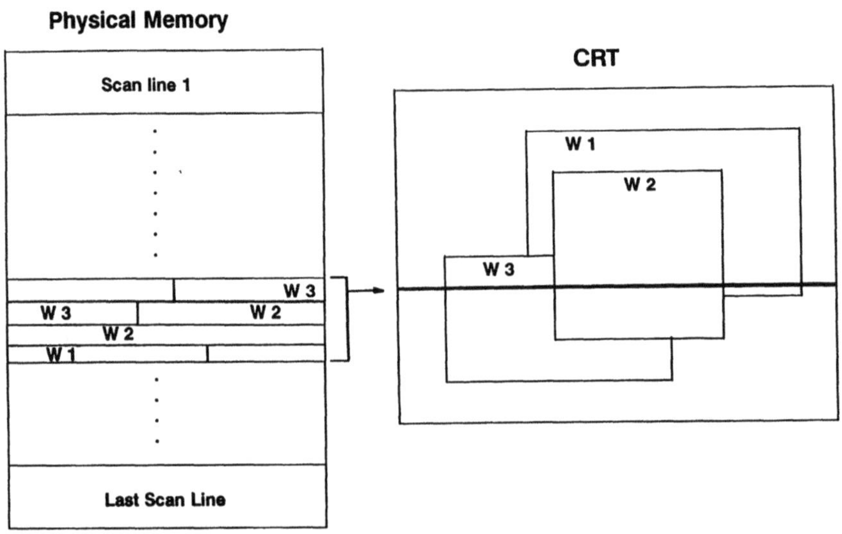

Figure 1: A Direct Mapped Graphics Memory

To address the shortcomings of direct memory mapping in general purpose computer systems, virtual memory systems were introduced. More recently the concept of virtual memory has been applied to graphics memory systems. Several commercially available VLSI chips [2,10] have support for segmented addressing schemes and a complete system using segmented graphics memory has been presented [7].

The use of segmented memory for graphics display sub-systems solves the performance problems inherent in the direct mapping systems. The segments are allocated so that each window has a continuous piece of physical memory of adequate size. Since the entire window is stored in the segment, the rendering engine does not need to clip pixels. The physical memory is large, allowing backing store for windows so that exposure events can be handled without disrupting the application processes and without additional rendering.

However, when segmented graphics memory is used, two new problem areas emerge. The first is the problem encountered in general-purpose CPU segmented memory systems which is also encountered in graphics segmented memory systems. This is complicated memory allocation and large-grained fragmentation. The problem of memory management is compounded as window resize operations are allowed. The second is the assembly problem unique to graphics. Assembling the multi-window display for viewing on the CRT requires accessing physically disjoint memory locations from different windows to display adjacent pixels. This is not feasible for high resolution displays with complete window capability. An example of a CRT display and the physical memory content for a segmented memory display is shown in Figure 2.

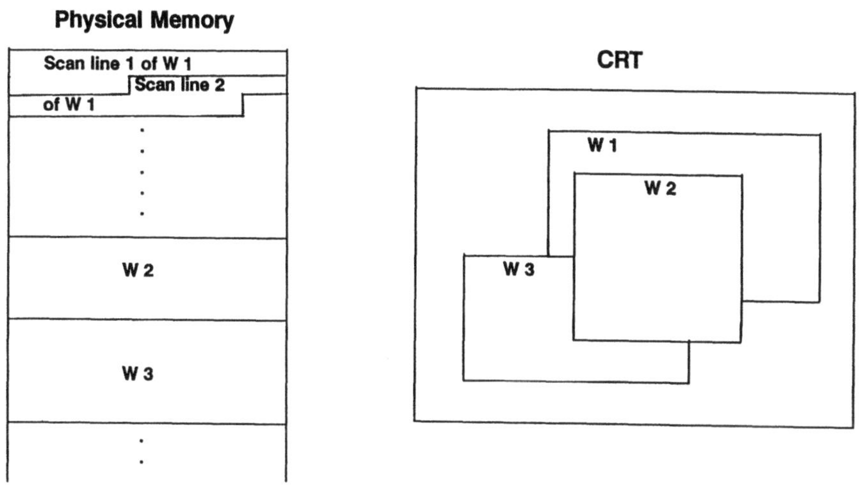

Figure 2: A Segmented Graphics Memory

To expand on the second segmentation problem it should be noted that a random access to physical memory using today's DRAM technology requires about 150 nS, while a high resolution CRT (1280 X 1024 at 60Hz non-interlaced) requires a pixel clock of under 10 nS. If a system allowed an arbitrary number of windows at arbitrary positions, then in the worst case a new window could start at each pixel location and so require a random access to physical memory at the pixel clock speed. Since physical memory can not be accessed at pixel clock speed, systems have been designed where windows must be placed on fixed boundaries [7] or the number of windows that may be displayed at one time are limited [10].

The problems with segmentation were solved in general purpose computer systems by using paged virtual memory systems. A design using a paged virtual memory system for the CPU, where the CPU also serves as the rendering engine, has been presented [1]. In this system graphics rendering used the same datapaths and memory provided for the CPU. Since it was critical to CPU performance, this memory system was designed for both CPU and graphics performance. A second memory was needed for CRT refresh.

Adding paged virtual memory to a dedicated graphics memory system solves the segmentation problems except for the assembly problem. The paging method has another graphics specific problem. The page organization is awkward for graphics. Shown in Figure 3, is a CRT display and the physical memory contents. As can be seen in the figure, Window 1 (W1) has a single scan line that is stored in physically disjoint pages. Note also that the data forming the scan line is not aligned to word boundaries within the page. This makes the control problem more complex when moving graphics data with a small outline (such as a character from a stored font) to an arbitrary location in a large window. This operation is also limited in speed by the inability to fully use the static column access mode of the DRAM chips to increase the transfer rate. The awkwardness of the paged system is also evident when a window resize operation is performed. As an example, if a window were to be expanded by one pixel on the left side, then the data for the extra pixel on each scan line would have to be inserted at the start of the data for each scan line. This would require reorganizing the data within all pages that form the window. This problem is a consequence of the physical memory organization not mapping to the requirements of windowed graphics.

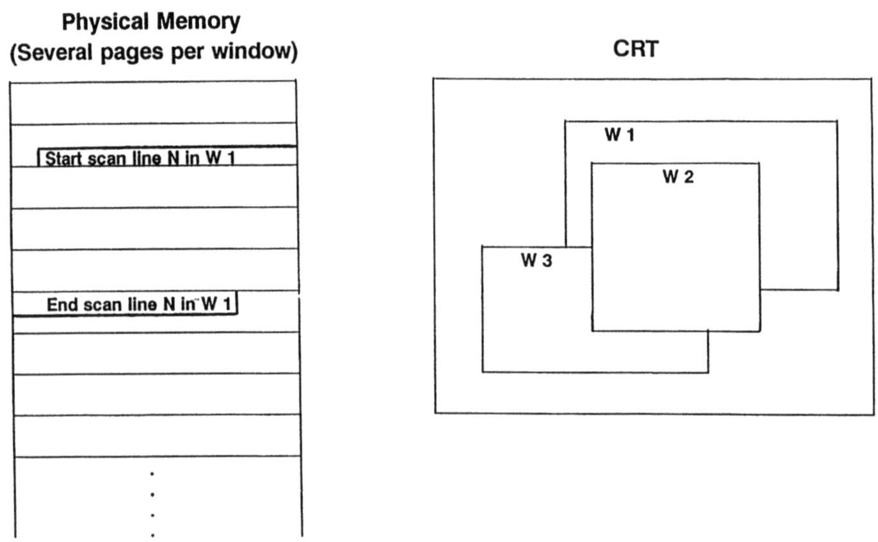

Figure 3: A Paged Graphics Memory

Therefore as graphics memories have developed along the same lines as those of general purpose CPU's, two important problems for graphics systems have not been solved. One is the problem of assembling a multi-window image for CRT refresh and the other is mapping virtual windows onto physical memory in a manner which facilitates high performance graphics operations.

Requirements for High Performance Rendering in the X Windows Environment

Users of modern computer graphics systems are starting to expect the ability to not only run, but to interact simultaneously with multiple applications [8]. At the same time users are requesting more graphics performance. The basic functional capabilities are being provided by building applications to run within window systems such as X Windows [9].

The VIEWS graphics memory sub-system was developed in conjunction with a high speed rendering engine that rendered over one million light-sourced triangles per second [3]. The system could also be used with other high performance rendering hardware [11]. The increased performance that users need is provided by building such advanced rendering hardware.

VIEWS was conceived to support this full rendering speed in an X Windows environment where window manipulations would appear instantaneous to the system user. To meet this design goal without compromise, the development process began with a list of system requirements. The list was broken into three parts: performance requirements, X Windows functional requirements, and advanced graphics feature requirements. This list became the basis for the VIEWS definition.

VIEWS Requirement List

1. PERFORMANCE REQUIREMENTS

— In order for the rendering hardware to run at full speed, window clipping had to be hidden from the rendering process.

— The memory organization should use the fast access mode (Static Column) of the DRAM chips whenever possible in rendering and data copy operations. This means the structure of the memory chips must be matched to the locality of graphics operations.

— Bandwidth requirements for data movement must be minimized. Specifically, pointers should be changed rather than copying data for large data movement operations such as switching buffers in double-buffering.

— The organization of data in the memory chips must allow the high speed sequential accesses required for CRT refresh.

— The method used to map pixel X,Y addresses to the physical memory array must not be so complicated that memory accesses become a performance bottleneck.

2. X WINDOWS FUNCTIONAL REQUIREMENTS

— The display system may not limit the number of windows that can be opened at once.

— The display system must allow windows to be placed at arbitrary locations on the display surface.

— Windows must be retained with effective utilization of available memory and graceful degradation when memory capacity is exceeded.

— Resizing of windows should appear instantaneous to the user.

— The backing store should be expandable.

3. ADVANCED GRAPHICS FEATURE REQUIREMENTS

— Double buffering must be supported.

— Panning within a window must be supported.

— Anti-aliasing must be supported by allowing several frames of the same scene to be stored and then combined to produce the final image.

— Video output should be flexible, to support several display formats. These would include, a high resolution CRT display, NTSC video, an as yet undefined HDTV format, and a stereo display.

— Beyond double buffering the backing store should support many buffers for storage and display of successive frames of animation sequences.

These requirements ruled out using a direct mapped frame buffer, so a virtual memory graphics system was investigated.

The first problems that needed to be solved were the memory organization and assembly problems. These problems are unique to supporting virtual memory in a graphics environment.

To solve the memory organization problems, a new virtual memory mapping scheme was developed. By extending the same mapping technique the assembly problem was solved and ALL the listed requirements for high performance rendering in an X Windows environment were met.

Graphics Virtual Memory to Physical Memory Mapping

The common concept in the three memory organizations already reviewed is the memory space is treated as a linear or 1-dimensional array. This is true in both the physical and virtual spaces. The key to solving the memory mapping problem was to recognize that graphics virtual space is inherently 2-dimensional. VIEWS maps a 2-D virtual address space on to a 1-D physical space. The virtual space is partitioned into rectangular regions, called *tiles,* that map into same-sized regions of the linear physical memory space. The size and shape of these tiles are fixed just as page size in a linear virtual memory system is fixed.

The transformation between the 2-D virtual space and the 1-D physical memory array is done as part of the address translation. In this system a 2-D virtual address is sent to translation hardware where the mapping is done by a combination of a translation RAM and a unique method of hardwiring the address lines through the translation table into the physical memory array. A schematic view of the virtual 2-D address field and its connection to physical memory is shown below as Figure 4. This translation hardware is analogous to the page translation hardware in a CPU with paged virtual memory.

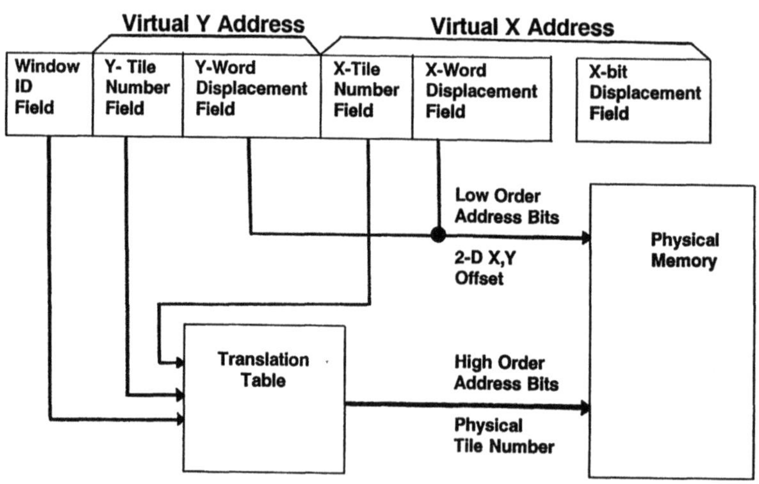

Figure 4: A 2 Dimensional Virtual Address Translated for Physical Memory

From the virtual address field it can be seen that virtual space is treated as a 2-D space, where the X bit displacement, X word displacement, and X tile number form one axis and the virtual Y address forms the other.

In order to go from virtual space to physical space, the virtual X tile number and virtual Y tile number are combined with the window ID to go through the translation table to form part of the physical address. The rest of the physical word address is formed by passing the X word displacement and Y word displacement directly to the memory array. (As is typical in graphics, the X bit displacement is used to build a mask that tells the graphics logic unit which bit to operate on within the word rather than to address a word.)

Note, the field alignment is altered in the translation process. The X word and Y word displacement fields from the virtual address are combined to form the low order field of the physical address. This re-alignment does several things:

- The tile is stored in a linear array of physical memory, but represents a 2-D region of virtual space.
- Data from the same rendering locality (pixels near each other in X and Y) are mapped onto a contiguous sequence of physical memory words. This allows fast static column accesses in the DRAM chips to pixels adjacent in either the X or Y directions.
- The width of a window is removed from the address calculation. In every previous method (direct, segmented, or paged), the data for adjacent scan lines is offset in linear address space by the length of a scan line in that window. In this 2-D mapping, the word offset in physical memory is based simply on the virtual X and Y offset of the pixel from its window origin.

An example of the mapping between the virtual space and the physical memory has been selected such that the 2-D virtual space will map to a Row Address Select (RAS) region in the linear DRAM address space. In other words the X word and Y word displacement fields will be combined to form the Column Address field going to the physical DRAM device. In Figure 5 a single tile from the 2-D graphics space is shown. Shaded in the figure are the pixels that would be stored in the same Row of a single DRAM device. Note that the bit position within the word is obtained from the X bit displacement field.

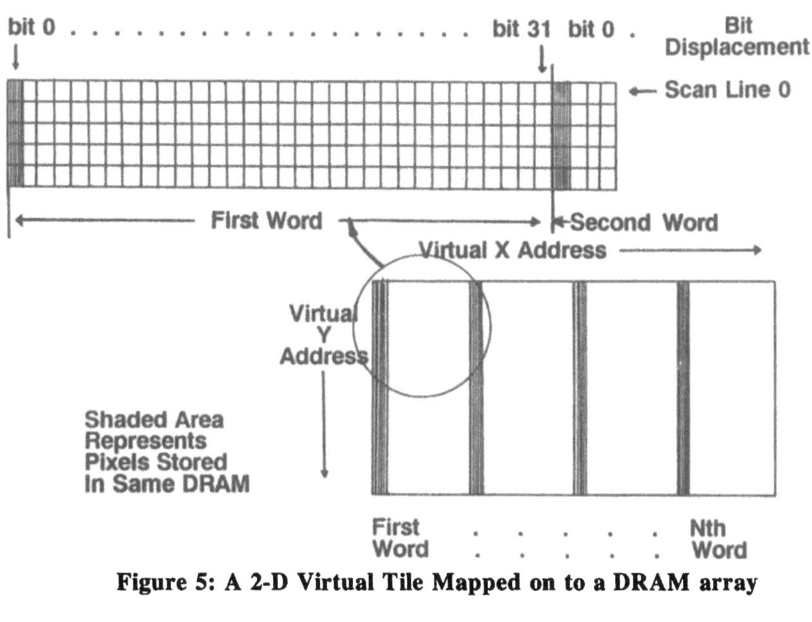

Figure 5: A 2-D Virtual Tile Mapped on to a DRAM array

Figure 5 showed the mapping of the bits within a 2-D tile to a localized region of the address space of the physical devices that form the memory array. Figure 6 expands on Figure 5 by showing how the example tile is held in the DRAM memory array.

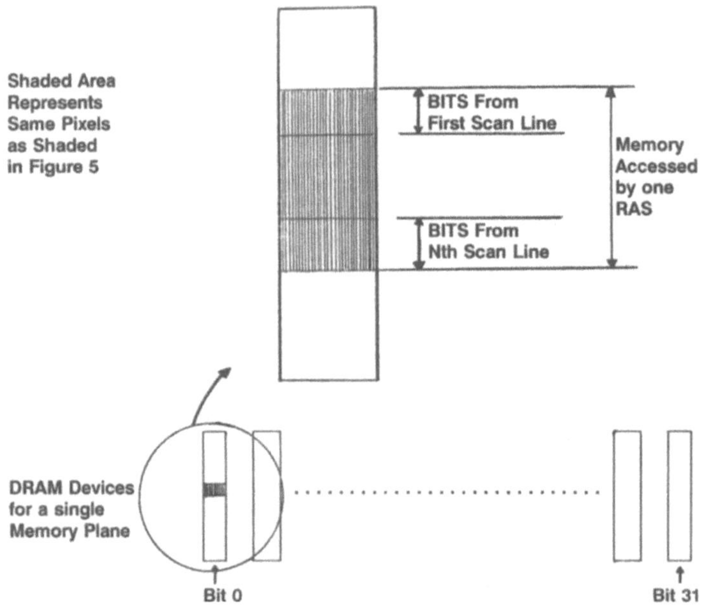

Figure 6: A Tile Stored in the Physical Memory Array

The term *tile* has been used in previous graphics memory system descriptions [5] to denote the array of pixels addressed simultaneously in the DRAM array. The VIEWS system labels this same array of pixels a memory *word*. Data for this group of pixels may be accessed in parallel in a single DRAM access time. In VIEWS this word corresponds to a 1 x 32 horizontal row of pixels in order to match the horizontal scan of both the CRT refresh and the rendering processes.

The VIEWS tile, in contrast, is a new concept analogous to a virtual memory page but with a 2-D organization. In Figure 5, the example 2-D Virtual Tile is the second level of the hierarchy of graphics locality and of DRAM access speed. This large tile will contain the typical small vector, triangle, raster character, or other graphics primitive. It can be accessed by a sequence of fast static column cycles without resorting to full DRAM RAS cycles.

These VIEWS tiles are then organized into virtual 2-D spaces called *canvas buffers* indexed by window ID. A canvas buffer contains all the tiles of a virtual window, whether displayed or not. When a window is opened, a set of physical tiles are allocated for its canvas buffer according to the size of the window. Then the translation table entries are updated to establish the virtual-to-physical mapping.

A schematic representation of a tiled memory array is shown in Figure 7. Window 2 (W2) has been shown filled with a crosshatched pattern on the CRT. The tiles that form the canvas buffer for window 2 are also shown with a crosshatched pattern. Note that the tiles forming the canvas buffer for window 2 may be disjoint in physical memory, just as pages for a linear address space may be disjoint in a conventional CPU with a paged memory system. Also note that within each tile there is a 2-D locality, just as a page has linear locality in the conventional system.

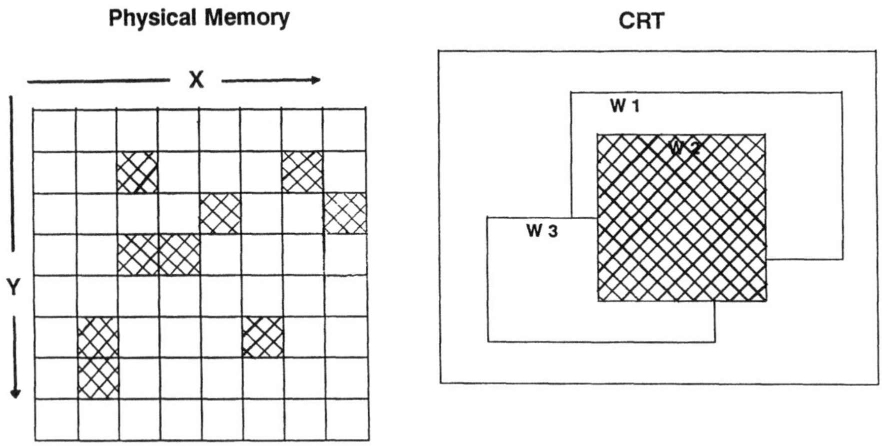

Figure 7: A Tiled Graphics Memory

Extending The Virtual Memory Mapping Concept

The virtual memory organization presented solves the virtual to physical memory mapping problem for graphics display systems. The second problem encountered with adding virtual memory to the graphics system is the data assembly problem.

The assembly problem has been previously addressed. The solutions include replacing the DRAMs used in the frame buffer with fast Static RAMs. Another is to use a separate physical memory to assemble the frame [1]. Also a system that avoids the assembly problem has been presented [12] with the frame store built out of full custom wafer scale memory circuits.

The VIEWS CRT image is located in a contiguous rectangular virtual canvas buffer like any other window. This special canvas buffer, the *scan buffer,* is the same size as the displayed image, with a direct mapping of the virtual pixel X,Y address to the CRT image X,Y location. The scan buffer's image is composed of sections of the various windows visible on the CRT, originating in their own canvas buffers. As any canvas buffer, the physical tiles comprising the scan buffer are scattered in available physical memory.

The traditional frame buffer has a physical mapping to the CRT. In contrast the scan buffer exists in virtual space. However the physical organization of the scan buffer is still critical to CRT refresh.

CRT refresh is accomplished by the sequential accessing of memory words across one row of a single physical tile using static column cycles, followed by a full RAS cycle to get to the next tile. Tile addresses are mapped from adjacent in virtual and CRT image space to disjoint in physical memory space. By requiring the use of DRAM rather than VRAM (dual-port video RAM), this approach necessitates a hardware cost saving. The large memory word and ability to use static column access mode result in adequate DRAM bandwidth for CRT refresh.

The scan buffer must be modified whenever windows are moved or double buffer swaps occur. If the windows happen to be tile-aligned, the scan buffer is modified by simply updating the translation table entries to map different physical tiles to the CRT image. In this case, the scan buffer requires no additional physical memory to be allocated for CRT refresh.

Windows are not typically tile-aligned. Therefore, tiles of the scan buffer containing non-aligned windows must be allocated an additional physical tile into which the correct data is copied from the appropriate canvas buffer(s). When windows are moved or buffers swapped in double-buffering, the affected scan buffer tiles need to be redrawn by data copying from the original canvas buffers.

When the contents of a window is updated but the window definition does not change, the contents of the updated canvas buffer can be used to update the scan buffer. The tile list that forms the scan buffer is left unchanged. Scan buffer tiles fully covered by an aligned window have already been updated in physical memory. In general, scan buffer tiles are recopied during CRT retrace from the canvas buffer being updated using a high-speed BITBLT capability.

This paper is concerned with the organization of a virtual memory system for bit-mapped graphics. As such, only a brief description of a system designed around the concept will be discussed. First, the functional model will be presented followed by a brief description of the hardware.

VIEWS Functional Model

The window system manager acts as a resource manager. This display sub-system is seen as rendering and display resources. These resources are:

- The Scan Buffer - This serves the function of a traditional frame buffer. The scan buffer is an image of the CRT screen built in a virtual graphics space. The display system micro-processor builds the scan buffer from the visible windows, stored in other tiles, and state information that tells the processor the window priority and screen location. Note that the screen location information could be a list of scan line endpoints for support of irregular shaped windows.
- The Canvas Buffers - These buffers store the entire rendered image for each window. The scan buffer is built by combining parts of several canvas buffers. The system design has two types of canvas buffers. These are the deep (24 bits of color and 4 bits for overlays) and shallow (2 bits deep) canvases. This is consistent with the X Windows scheme for bitmap allocation. The shallow bitmaps are seen as storing alpha-numeric windows and glyphs [fonts].
- Color Look Up Tables were supported.
- The last resource that the window manger has access to in this system is the general purpose micro-coded engine. This processor will process high level commands that the window manger sends. These commands would include window manipulation tasks, and some rendering tasks. The rendering task that would be handled are alpha-numeric updates, some graphics updates and cursor updates.

There will still be cases where the opened windows require more tiles than the system has available. When this happens, tiles can be taken from a window not being updated and used for the current window being rendered. It is important to note that once the visible part of a window is placed in the scan buffer, removing the canvas buffer will not affect the display. Also note, if the scan buffer is then changed and the window that had its canvas buffer removed is now more exposed, then an exposure event will be generated and sent back to the application following the standard X protocol. In this case, the system performance will degrade gracefully, and instantaneous window manipulation will no longer be possible.

VIEWS Hardware Model

Shown in Figure 8 is the high level block diagram of the VIEWS system.

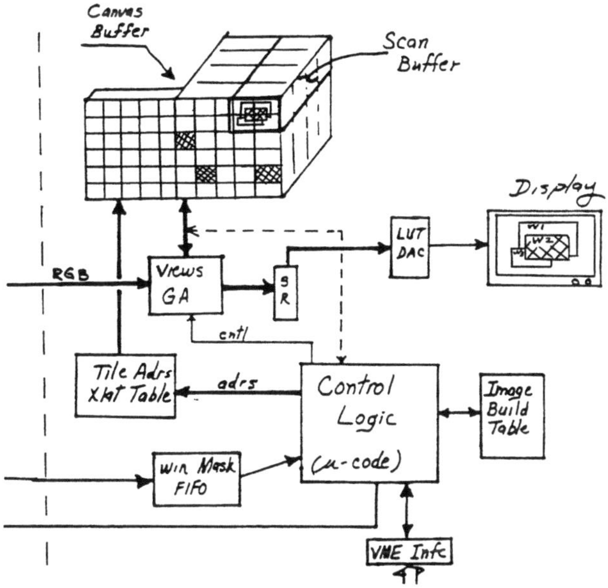

Figure 8: VIEWS Block Diagram

The lines on the left side of Figure 8 show the pixel data and Z status bit entering the VIEWS systems from the supported rendering engine [3].

The VIEWS gate array is an ASIC device that serves as the focal point of data transfers within the VIEWS system. The gate array caches the pixel data [5] from the rendering engine, and is used to store data during high speed data transfers within the memory array (BITBLT). A display FIFO within the VIEWS gate array allows static column mode accesses to be used for reading the tiles that form the scan buffer during CRT updates. The contents of the display FIFO are sequentiality output to the high speed logic section. One VIEWS gate array is used per memory plane. The connection between the gate array and the memory array is 32 bits per plane. This datapath can support a maximum burst transfer rate of 1.12 G bytes per second. There is more than enough bandwidth to allow a full screen (1536 X 1152 pixels) image to be moved within the memory array and the CRT to be updated within one frame time of 16.67 mS.

The Z win mask FIFO caches the Z win bits. These bits are used to condition the control signals going to the VIEWS Gate Array.

The control logic contains all the address generation circuitry, the CRT refresh control circuitry and the micro-coded processor. The micro-processor is responsible for controlling the VIEWS environment, including processing commands from the window manager, initiating and synchronizing VIEWS hardware operations, managing data structures, and performing some low level rendering tasks.

The image build table is the table in the processor's RAM space that defines the way the scan buffer is built.

A simple static ram is used for the tile address translation table. The translation table does not supply status information, such as a valid bit. Instead, the u-processor that controls the sub-system has to keep track of the status of the memory array. Part of the control process is to make sure that before the rendering engine starts to update a window all the required tiles are present in memory. This requirement assures that tile faults can not occur.

The memory array design supports several physical memory array sizes. The maximum physical memory allowed by the design is 4K X 4K pixels in the deep canvas buffers and scan buffer combined. The shallow buffers can be expanded to 8K X 4K pixels. Note that these limitations are due to the VIEWS implementation and are not architectural limits.

The high speed logic includes ECL shift registers and a LUT/DAC used to drive a high resolution (1536 X 1152 pixels) CRT display.

In VIEWS, the word size is 32 bits, so the bit displacement field is 5 bits long. The tile size is 256 X 256 pixels. The X word displacement field is 3 bits (8 words X 32 pixels/word = 256 pixels). The Y word displacement field is 8 bits. The virtual space is limited to 4096 X 4096 pixels so the X tile displacement and Y tile displacement fields are 4 bits each. The Window ID field is 8 bits long.

The memory array is designed using DRAMs and not VRAMs. The DRAMs cost less and are more flexible on the output port than VRAMs. The control logic allows the refresh to be programmed to any resolution (up to 1536 X 1152 pixels at 60Hz). Some of the programmed resolutions excepted to be used are the very high resolution mode, a high resolution mode (1280 X 1024), and NTSC video. About 1/3 of the high speed BITBLT bandwidth is allocated for the display update task.

Extensions and Limitations to Current System

The system as presented is designed to replace current high end display sub-systems. The architecture would not be cost effective in today's low end graphics systems. A low end implementation of this system would cost about the same as current high end double buffered display sub-systems.

In a low end implementation about half the memory would be needed by the scan buffer, leaving one screen sized piece of physical memory for the deep canvas buffers. An area of several screen sizes, could also be included, for shallow canvas buffers, without adding much cost. Such a system configuration would offer the full functional capabilities of VIEWS, but may not deliver all the performance the architecture can provide, since exposure events would be sent back to the application programs often. One big advantage to using the VIEWS architecture over current designs is that as the next generation of DRAMs become available the memory chips could be changed, while all of the control hardware and software remain untouched. At that time, the physical memory would hold a scan buffer with tiles left over for 3 full screen sized deep canvas buffers.

As new generations of memory chips become commodity items, the price per bit is reduced. The VIEWS memory organization is well suited to take advantage of the evolving DRAM technology. With proper care, a design today would not only plan for the next generation of DRAMs, but allow for many future generations of memory devices.

As larger memory arrays become economically feasible, the simple static ram translation table in VIEWS could be replaced by the more sophisticated translation means used in general purpose virtual memory CPU designs.

Another interesting extension to investigate would be sharing a large memory array between the CPU and graphics memories. This has been done using a linear mapping for both memories [1]. However the extension to this architecture would require a process could request the physical memory map to a either a 2-D or linear virtual space on a process by process basis. In other words both *tiles* and *pages* would be held in the same physical memory array.

Conclusion

A system has been presented, that meets ALL the stated design goals. The system allows high performance graphics in an environment that consists of a hierarchy of resizable windows. By providing instantaneous response to most window operations, the user can feel that he is interacting simultaneously with several applications.

Industry standards and state-of-the-art performance are not mutually exclusive! VIEWS is the first system designed to synergistically combine the industry standard X Window system and state-of-the-art rendering technology for unparalleled interactive graphics.

Acknowledgements

The author would like Jorge Lach to be recognized as co-inventor of the system described. Thanked are Tom Durant for his work on the implementation details, Michael Deering for his review of the system, and Roger Day for both his leadership and moral support. Efforts by the rest of the Applicon engineering crew are greatly appreciated.

The Hewlett Packard Graphics Technology group is thanked for supporting and reviewing this paper, especially Gary Taylor.

REFERENCES

1. **Apgar, Brian et al,** "A Display System for the Stellar Graphics Supercomputer Model GS1000", *Computer Graphics,* **22,** 4, 1988,(Proc SIGGRAPH)

2. **Asal, Mike et al,** "The Texas Instruments 34010 Graphics System Processor", *IEEE CG&A* **6,** 10, (October 1986)

3. **Deering, M. et al,** "The Triangle Processor and Normal Vector Shader: a VLSI System for High Performance Graphics", *Computer Graphics,* **22,** 4, 1988,(Proc SIGGRAPH)

4. **Foley, J. D. and Van Dam, A.,** "Fundamentals of Interactive Computer Graphics", *Addison-Wesley,* 1982

5. **Goris, Andy et al,** "A Configurable Pixel Cache for Fast Image Generation", *IEEE CG&A* **7,** 3

6. **Hamacher, Carl V. et al,** "Computer Organization", *Second Edition McGraw-Hill* 1984

7. **Ilgen, S. and Scherson, I.D.,** "Real Time Virtual Window Management for Bit Mapped Raster Graphics", *Proc. 5th International Conf. on Computer Graphics in Japan, Springer-Verlag, Tokyo* 1987

8. **Lantz, Keith A. et al,** "Reference Models, Window Systems, and Concurrency", *Computer Graphics,* **21,** 2, (April 1987)

9. **Scheifler, Robert W. and Gettys Jim,** "The X Window System", *ACM Trans. Graph.* **5,** 2 (April 1986)

10. **Shires, Glen,** "A New VLSI Graphics Coprocesssor - The Intel 82786", *IEEE CG&A* **6,** 10, (October 1986)

11. **Swanson, Roger and Thayer, Larry,** "A Fast Shaded-Polygon Rnederer", *Computer Graphics,* **20,** 4, 1986,(Proc SIGGRAPH)

12. **Westmore, Richard J.,** "A Window-Based Graphics Frame Store Architecture", *ACM Trans. Graph.* **7,** 4 (October 1987)

A Real-Time Raster Scan Display for 3-D Graphics

D. Jackèl
TU Berlin, Institut für Technische Informatik
Franklinstrasse 28–29, D-1000 Berlin 10

and

H. Günther, B. Herwig, H. Rüsseler
GMD-FIRST an der TU Berlin
Hardenbergplatz 2, D-1000 Berlin 12, FRG

Abstract

This paper describes the architecture of a raster scan display for real-time visualisation of *shaded* polygons. A performance of $15 \cdot 10^6$ Phong shaded pixels per second is a primary goal of a pipelined rendering processor. The performance of the geometry processor, which is responsible for the geometrical transformations, the 3-d clipping and the perspective projection, will exceed 100,000 triangle shaped polygons.

Following a survey of the entire 3-d real-time system, we will describe architectural details of the rendering processor. Finally, the main features enabled by the architecture are highlighted.

1. Introduction

The hardware of conventional raster scan displays supports only the generation of 2-d graphical primitives (*e.g.* points, vectors, circles, etc.) and their mapping from object-space to screen-space. The 3-d visualisation processes, which are extremly time consuming, are executed by the general purpose processor(s) of a host-system.

To fulfill the requirements for 3-d real-time visualisation, high performance graphic architectures have changed significantly. The characteristic of such architectures is the hardware supported viewing pipeline, for which a variety of architectural solutions have been introduced in the past decade.

In order to classify these architectures, we have to divide the visualisation process sequence into its geometry processes and into its *rendering* processes. All vertex oriented processes (*e.g.* geometrical transformations, 3-d clipping and perspective projection of a vertex) belong to the class of geometry processes. All pixel oriented processes (e.g. determination of coordinates and rgb-values of a pixel) belong to the class of rendering processes. These classifications help us to differentiate four basic architectural structures:

- Both the geometry processes and the rendering processes are supported by pipeline architectures. The architectural concept of the Silicon Graphics IRIS-Systems [1] as well as the Hewlett Packard SRX graphics engines [2] are representative for this type of graphics system.

- The geometry processes are supported by a pipeline architecture and the rendering processes are supported by a parallel architecture. An example of this are the Pixel Machines of AT & T described by Potmesil and Hoffert [3].

- The geometry processes are supported by a parallel architecture and the rendering processes are supported by pipeline structured architecture. The TITAN of ARDENT and the STELLAR of STELLAR-Systems described by Diede et al [4] and [5] are representative for this type of architectural structure.

- The support for the geometry processes, and the rendering processes are supplied by a parallel structured architecture, which can be found in the PIXAR-System described by Levinthal and Porter [6].

Our system introduced in the following belongs to the third group. The objectives for the capability and performance of the system are:

More than 100,000 triangular polygons should be processed per second. The geometry processes include: rotation, translation, scaling, backfacing, 3-d clipping and perspective projection.

$15 \cdot 10^6$ *pixels* should be rendered per second. The rendering process includes: calculating pixel coordinates, computation of the pixel normals, Phong-shading, HSL to RGB-conversion and hidden surface removal.

At first, a survey of the graphic system and its basic function will be presented, and some implementation details of the rendering processor will be discussed. This is followed by a description of the display file structure and the synchronisation of the geometry processors. Finally, we will summerize the main features and present the state of implementation, including the planned extension and improvements of our system.

2. System Survey and Implementation

The main subsystems of the real-time display, as shown in Figure 1, are :

a geometry subsystem
a rendering subsystem
a frame-buffer subsystem
a 2d-subsystem.

In the following, we will describe the tasks, the basic principles, and the architecture of these subsystems.

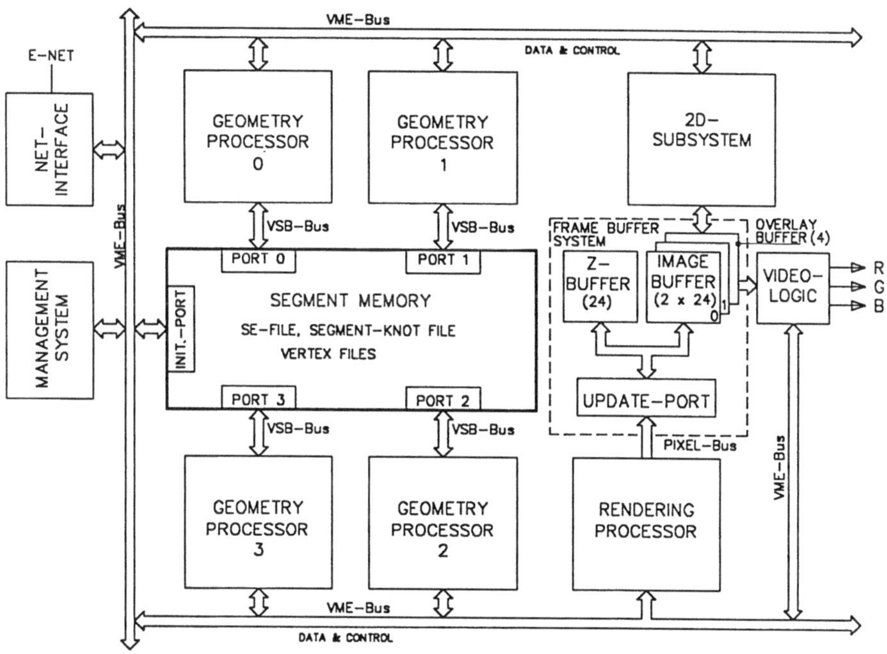

Figure 1: Architecture of the Real-Time Raster Scan Display

2.1 Geometry Subsystem

The main objective of the geometry processing unit is the efficient handling of geometrical rotation, translation and scaling of 3-d objects. Moreover, this subsystem executes the *backfacing*, *3-d clipping* and the *perspective-projection* processes. The pipeline of the geometry processes terminates after computing the *parameter set* needed for the initialization of the *rendering processor*. All geometry processes mentioned above are executed in real time for more than 10,000 triangular polygons every 0.1 second by using four processing units working in parallel.

For the initialization of the rendering processor, two methods are applied to generate the parameter set mentioned above.

Using the first method, the processing pipeline will be started via function calls. Therefore, a set of graphical routines are implemented.

The second method for generating the parameter set implements a display file, referred to in this paper as *object-descriptive data structure* (*ODDS*). Although this method is not as flexible as the first one, it is more efficient. It will be used mainly if a model of a 3-d scene has only fixed vertex coordinates. The *ODDS*-architecture and how this display file is processed by the geometry processor array are discussed later.

2.2 Rendering Subsystem

The execution of the *scan converting* process, which includes *shading* and *z-buffering*, is the main task of the *rendering processor*. The components of the rendering subsystem are shown in Figure 2.

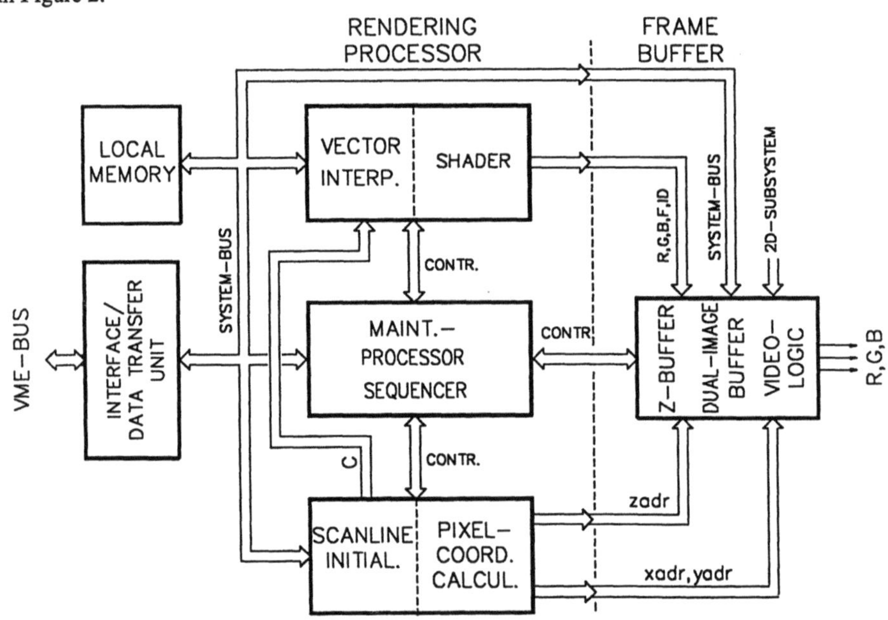

Figure 2: Block Diagram of the Rendering Subsystem

At the core of the rendering subsystem are two pipelines. The task of the first pipeline, consisting of a vector interpolator and a shading unit, is to calculate the rgb-values of the pixels using the Phong illumination model. The second pipeline, with the functional units *scanline initialiser* and *pixel coordinate calculator*, is responsible for the determination of the z-coordinate values as a function of the xy-coordinates.

In addition to the pipeline units, a local memory, a maintenance processor, a VME-bus interface, and a data-transfer unit make up the rendering subsystem. The tasks of these units are discussed in short at the end of section 2.2.

2.2.1 Rendering Pipelines

parameter set. As mentioned in section 2.1, the geometry subsystem generates parameter sets for the initialization of the rendering pipelines. Each parameter set describes a triangle-shaped polygon, which decomposes into a unified form as shown in Figure 3.

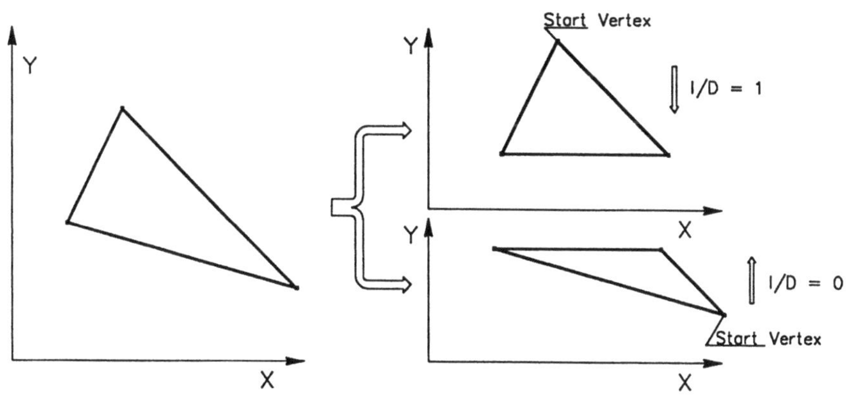

Figure 3: Decomposition of the Triangles into Unified Forms

The decomposition process is necessary to achieve a simple parametrical description of the polygons. By this measure, we obtain a more efficient hardware solution for the rendering pipelines.

Figure 4 shows the organisation of the parameter sets for *smooth-shaded* and *constant-shaded* polygons.

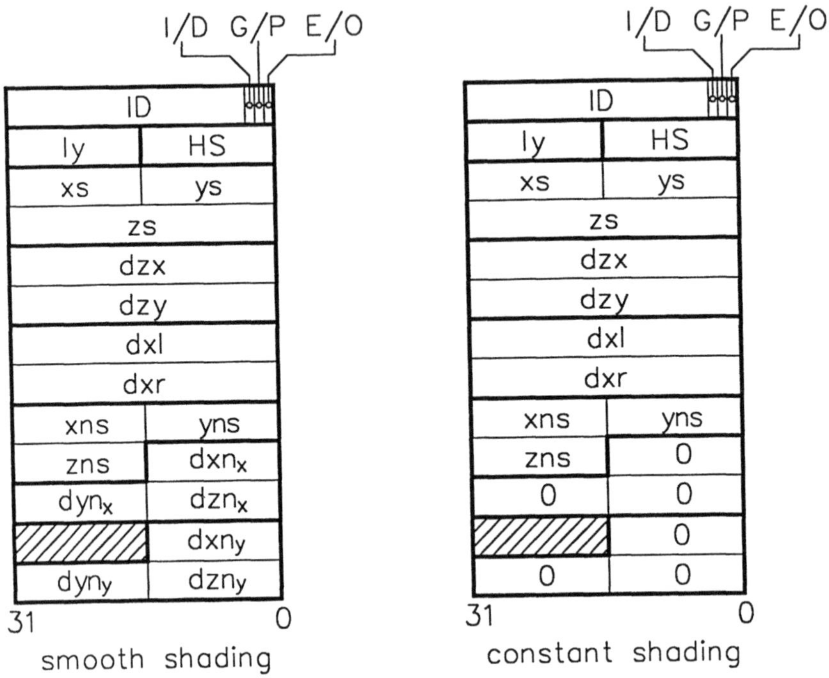

Figure 4: Parameter Sets for Initialising the Rendering Pipelines

The functions of the parameters are discussed in the following:

ID: The ID is a concatenation of the segment identifier SID and the surface element identifier SEID and is used for man-machine-interactions.

I/D: The I/D-bit indicates a decrementation (I/D = 1) or an incrementation (I/D = 0) of the y-coordinate by the scanline generation process.

G/P: The G/P-bit initializes the shading pipeline for executing either the Gouraud or the Phong shading process. For G/P = 0 the shader is switched into the Gouraud shading mode. Otherwise (G/P = 1), the Phong shading mode will be selected.

E/O: The E/O-bit controls the dual image buffer. E/O = 1 selects the odd and E/O = 0 selects the even image buffer unit.

ly: ly is the y-address of the last scanline which is rendered by the pipeline units (see Fig. 5a).

xs,ys,zs: [xs ys zs] is the vertex position from which the rendering process starts (see Fig. 5a).

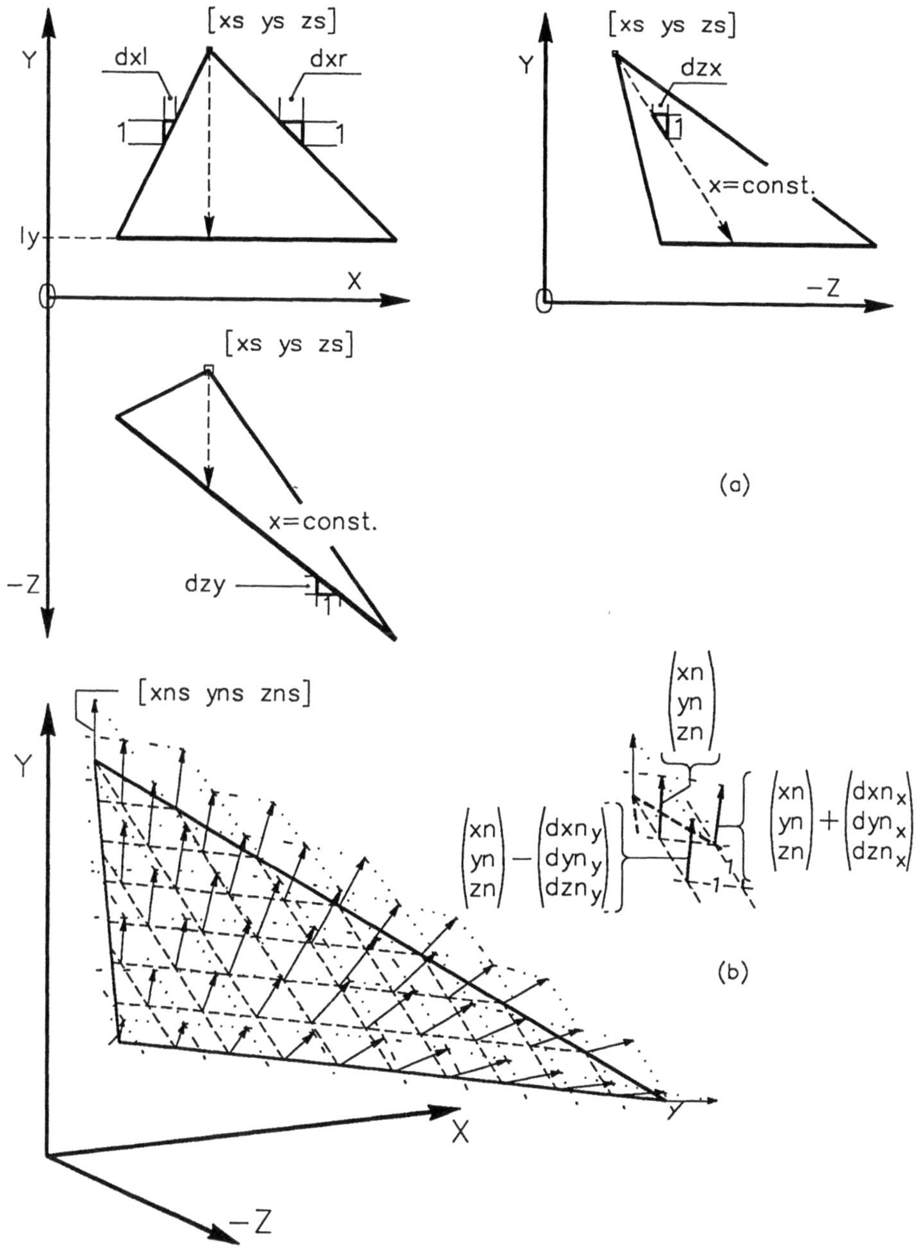

Figure 5a,b: Function of the Parameter: (a) ly, xs, ys, zs, dzx, dzy, dxl, dxr (b) xns, yns, zns, dxn_x, dyn_x, dzn_x, dxn_y, dyn_y, dzny

dzx,dzy: dzx and *dzy* represent the z-slope increment of the SE-plane. *dzx* is a function of an x-incremental step; *dzy* is a function of an y-incremental step (see Figure 5a).

dxl,dxr: dxl and *dxr* represent slope increments of the left and right SE-edges respectively. Both are functions of an y-incremental step (see Figure 5a).

xns,yns,zns: [xns,yns,zns] is the vertex vector from which the rendering process starts (see Figure 5b).

dxn_x, dyn_x, dzn_x: The vector $[dxn_x\ dyn_x\ dzn_x]$ is added to the pixel normal vector [xn yn zn] after executing an x-incremental step. For constant shading all elements of this vector are set to zero (see Figure 5b).

dxn_y, dyn_y, dzn_y: The vector $[dxn_y\ dyn_y\ dzn_y]$ is added to the pixel normal vector [xn yn zn] after executing an y-incremental step. For constant shading all elements of this vector are set to zero (see Figure 5b).

rendering algorithm. Before going into the details of the rendering hardware, the algorithmic structure of the process, which is supported by the pipelines, will be outlined.

At the beginning of the process, the initializing register of the rendering pipelines are loaded with the parameter set. Because the first scanline consists of only one pixel, no pixel interpolation is necessary. The steps b), c) and d) of the inner loop are executed directly.

From the second to n-th scanline the rendering hardware functions as follows:

First, all values needed for initializing the rendering process of one scanline, are computed. The outer loop of Algorithm 1 is responsible for the execution of this initializing process. Within this iteration the x-coordinate components x_l and x_r on both scanline vertices are determined. Additionally, the z-value *zl* and the components pixel normal vector *[xnl ynl znl]* must be calculated only on the left scanline vertex.

xl and *xr* are given by:

$$xl := xl' + dxl\ ;\ xr := xr' + dxr\ . \tag{1}$$

xl' and *xr'* are the coordinate components of the left and right vertices of the preceding scanline, respectively.

The z-value zl and the components of the pixel normal vector are given by:

$$zl := zm + (xl - xs) \cdot dzx \tag{2}$$

$$xnl := xnm + (xl - xs) \cdot dxnx \tag{3a}$$

$$ynl := ynm + (xl - xs) \cdot dynx \tag{3b}$$

$$znl := znm + (xl - xs) \cdot dznx \tag{3c}$$

zm, xnm, ynm and *znm* which are initalized by

zm := zs; xnm := xns; ynm := yns; znm := zns

are incrementally decreased or increased by each y-step:

zm := zm + dzy; xnm := xnm + dxny
ynm := ynm + dyny; znm := znm + dzny

Initializing the rendering processor by means of the precalculated parameter set:

if I/D = 0 **then** yinc := -1 **else** yinc := 1

Render the pixel at start vertex [xs ys zs] by the execution of steps b), c) and d) within the inner loop:

for y := ys + yinc **to** ly **step** yinc

Computation of the x-coordinate values x_l and x_r at the left and right vertex of the scanline:
Computation of the pixel normal [xnl ynl znl] at the left vertex: of the scanline

for x = x_l **to** x_r **step** 1

a) computation of the z-value
b) computation of the pixel vector [xn yn zn]
c) computation of the pixel lightness 'l'
d) Z-buffer operation and data transfer of rgb-values to the image buffer

next x
next y

Algorithm 1: Structure of the Hardware Supported Rendering Process

a and b) The scanline rendering is executed within the inner loop. For each x, within the interval *[xl xr]*, the z-component and the pixel normal vector *[xn yn zn]* is calculated (y =const.).

After allocating the initializing values of the scanline

$$z := zl, xn := xnl, yn := ynl \text{ and } zn := znl$$

all further z-values and normal vector components are given by:

$$z := z' + dzx \quad (4)$$
$$xn := xn' + dxnx \quad (5a)$$
$$yn := yn' + dynx \quad (5b)$$
$$zn := zn' + dznx \quad (5c)$$

z' and the vector *[xn' yn' zn']* are preceding values.

c) This is followed by the shading process. In order to execute this process, three main steps are necessary.

First, the result of dividing the *[xn yn zn]*-vector by the value *zn* is the unified form *[xn/zn yn/zn 1]*.

Second, the xn/zn- and yn/zn-components may be regarded as address pointers of a table memory. This memory contains the precalculated values of a light reflectance map (see [7]). Under these conditions, it is allowed to consider the light reflectance value *l* of a pixel only as a function of both parts of the address pointer *l = f(xn/zn, yn/zn)*.

In the third step of the shading process, the value *l* is concatenated with the hue *h* and colour saturation values *s* allocated to the surface element. For each of the 2^{17} *hs/l*-combinations the corresponding rgb-values can be allocated by a look-up table.

d) The final step of the rendering process is the hidden surface removal via z-buffering. Because this method is easily and efficiently supported by simple hardware, z-buffering is commonly used in contemporary real-time raster scan displays.

2.2.2 Hardware Organisation of the Rendering Pipelines

When initialising the rendering processor, all different *SE-parameters* are distributed to their own *FIFO*-memories. The *FIFO*'s are useful in achieving a sufficient load balance between the rendering and the geometry processor. Moreover, the access processes of both subsystems do not have to be synchronized.

As the *SE-parameters* are distributed to private *FIFO's*, the rendering pipelines can be initialized by one load cycle. In the following the hardware organisation of both rendering piplines will be discussed.

vector interpolation unit. The task of the *vector interpolation unit* is to determine the barycentrical-oriented pixel normal *[xnl ynl znl]* at the left vertex of the scanline with respect to the equations 3a-c. This process will be executed with four pipeline stages mainly consisting of six adders and three multipliers.

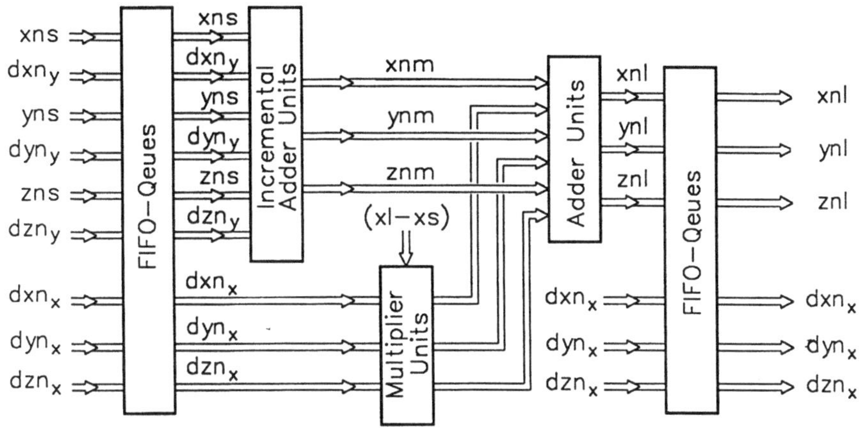

Figure 6: Blockdiagram of the Vector Interpolator

All values of the pixel normals are written into a second FIFO-bank. This functional unit operates as an interface serving as an asynchronous link to the shading unit and the vector interpolator, and ensures a sufficient load balance.

shader. The determination of the lightness values of the pixels within a scanline, is the task of the shader.

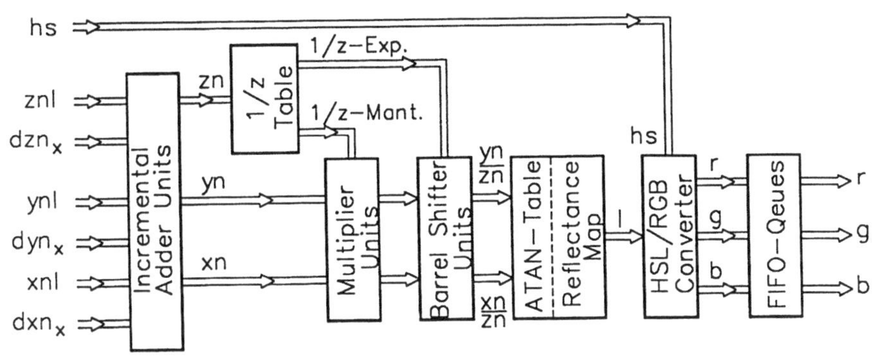

Figure 7: Block-Diagram of the Shader

In the first stage, the shading unit manages the interpolation of the pixel normals, corresponding to the equations 5a-c. The interpolation process is applied by using three incremental adders.

The next three stages are necessary to transform the pixel normal vectors *[xn yn zn]* into the form *[xn/zn yn/zn 1]*. To avoid the time consuming division by zn, a *1/zn*-table is used. The *zn-to-1/zn* table conversion is followed by a multplication stage. Both products xn•(1/zn) and yn•(1/zn) must be normalized. This is done in the fourth stage by barrel-shifting units.

In the sixth and seventh stage the determination of the *l-values* is easily accomplished by using a precalculated *reflection map*. Mainly under the restriction of *fixed-positioned virtual light sources*, the determination of a *l-value* only requires one access operation to the *reflection map* formed by a look-ahead-table memory. For precalculation of the 32,400 reflection-map values the *Phong Light-Model Equation* is applied.

The task of the last shader stage is the execution of the HSL/RGB-conversion. This is also accomplished by using three *look-up tables* (HSL/R, HSL/G and HSL/B) which have each a size of 128K x 8 bit.

scanline initializing. Scan-conversion is the task of the second pipeline, which is divided into a *scanline initializing unit* and a *pixel coordinate calculator*.

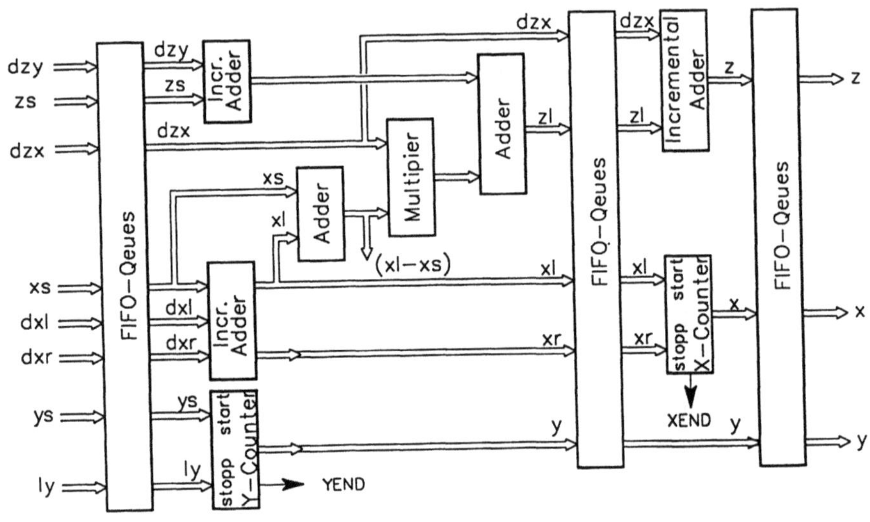

Figure 8: Block-Diagram of the Scanline Initializing and the Pixel Coordinate Calculator

In the first stage of the *scanline initializing* process two incremental adders are used to determine the *xl* and *xr* coordinate components according to the equations 1a and 1b. A third adder-unit is required to increment or to decrement the constant *dzy*-value to the old z-coordinate. The second stage is responsible for the determination of the difference *(xl-xs)*. The task of the third and fourth stages of the *scanline initializing* process is the determination of the *zl*-value according to the equation 2.

Moreover, the *scanline initializing unit* contains a binary y-counter. A comparator unit is used for the determination of the y-coordinate value and for indicating the last scanline by generating the *YEND* signal.

Analogous to the vector interpolator unit, all output values of the *scanline initializing unit* are written into the second *FIFO*-memory bank.

pixel coordinate calculator. The *pixel coordinate calculator* computes all *z-values* within a scanline (equ. 4), which are used for *hidden pixel removal* by means of *z-buffering*. The x-coordinate component is determined by a binary counter starting at the x_l*-value* and stopping at the x_r*-value*.

2.3 Frame Buffer Subsystem

The frame buffer supports a screen resolution of 1024 x 1280 pixels. As shown in Figure 6, the entire system consist of a 2 x 24-bit *dual image buffer*, a 24-bit *z-buffer* with an update port, a 4-bit *overlay-buffer*, a bank address computation unit and a buffered *I/O-multiplexer*. The *serial converter & multiplexer*, which works in parallel to the *I/O-multiplexer*, can be considered to be part of the dual image buffer. In the following we will describe these functional units in more detail.

z-Buffer. In order to achieve update cycles of 75 ns, the *z-buffer* implements the *complex memory interleaving* method. This ensures the rendering rate of $15 \cdot 10^6$ pixel per second, as required.

The z-buffer is partitioned into 20 independent-addressable memory banks. The mapping of the logical frame-buffer address to the physical bank-address is the task of the *bank address computation unit*. Moreover, this unit is responsible for bank addressing fault detection. Details of the *complex memory interleaving* method applied to frame-buffer systems can be found in Rüsseler et al [8].

The presetting of the z-buffer is necessary after each frame-cycle. Using *video-RAM's*, the preset time needs 240 ns for a group of four pixel rows, *i.e.* 0.7 ms for the entire z-buffer.

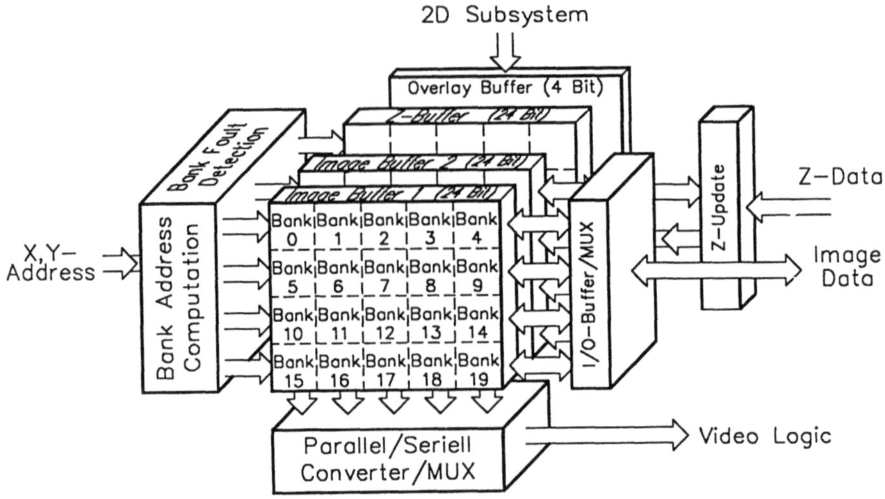

Figure 9: Frame Buffer System

dual image buffer. The dual image buffer, divided into an *even* and an *odd* unit, receives the *rgb*-values determined by the *shading unit.* For real-colour display, each r, g and b-component has a size of 8 bits. Analogous to the *z-buffer*, this functional unit is partitioned into an equal number of memory banks using the same memory interleaving method. The *'I/O-Buffer & Multiplexer'*, controlled by *E/O-bit*, is responsible for the parallel-I/O of the *rgb*-values to both image buffer units.

overlay buffer. The task of the *overlay buffer* is the display of alphanumeric information or vectorized graphical primitives. Because the *2-d subsystem* has the access right to this unit, it can work in parallel to the image-buffer and z-buffer units.

2.4 The 2-D Subsystem

The main objective of the *2-d subsystem* is the generation of 2-d primitives, for example, dots, 2-d vectors, circles, arcs, ellipses and alpha-numeric characters. Moreover, it is planned to use this unit for supporting the *man-machine-interaction* process based on the *X Window System.*

For the implementation of the 2-d subsystem an *off-the-shelf* graphics processor is applied.

3. Summary

We have presented an architecture for a 3-d graphics system for real-time polygon rendering. The main features of the system are summarized as follows:

100,000 triangle shaped polygons per second are processed by geometry transforms, backfacing, 3-d clipping and perspective projection.

$15 \cdot 10^6$ pixels per second are Phong-shaded and z-buffered by two rendering pipelines.

The size of the frame buffer is 1280 x 1024 pixels, with a total number of 76 bits per pixel.

Developmental status as of summer 1989: The pre-version of the geometry processor, the segment memory and the 2-d display is being completed. The frame-buffer system is being tested. The rendering processor is in the layout design state.

At the end of the year, we expect an operational prototype.

4. References

[1] K. Akeley, T. Jeremoluk : *High Perfomance Polygon Rendering*; Computer Graphics, Volume 22, No 4; pp. 239-246, 1988.

[2] D. Burgoon: *Pipelined Graphics Engine Speeds 3-D Image Control*; Electronic Design, July; pp. 113-119, 1987.

[3] M. Potmesil, E.M. Hoffert: *The Pixel Machine: A Parallel Image Computer*; Computer Graphics, Volume 23, No 3; pp. 69-78, 1989.

[4] T. Diede et. al.: The Titan Graphics Supercomputer Architecture; Computer, September; pp. 13-29, 1988.

[5] B. Apgar et al.: *A Display System for the Stellar Graphics Supercomputer GS1000;* Computer Graphics, Volume 22, No 4; pp. 255-262, 1988.

[6] A. Levinthal, T. Porter: *Chap - A SIMD Graphics Processor*; Computer Graphics, Volume 18, No 3; pp. 77-82, 1984.

[7] B.K.P. Horn: *Understanding Image Intensities*; Artificial Intelligence Vol. 8, No. 2; pp. 201-231, 1977.

[8] H. Rüsseler, H. Günther, D. Jackèl: *Eine Bildspeicherarchitektur für Raster-Displays mit Echtzeiteigenschaften*; ITG-Fachbericht 102; VDE-Verlag GmbH; 1988.

The Graphics Unit of the INTEL I80860

Ulrich Kursawe

KONTRON ELEKTRONIK GmbH
Research & Development Department
Breslauer Str. 2, D-8057 Eching, FRG

The Intel I80860 is a very powerful RISC processor, designed for applications that require a large amount of floating point and integer calculations. Additionally it supports graphics applications with a Graphics hardware unit. The aim of this article is to investigate, for which application this unit is useful and whether the results obtained by the help of this unit are better as with standard C or assembly implementations of the same algorithm.

1. General Aspects

Although the Intel I80860 is mainly designed as a fast floating point number cruncher, it also features a graphics unit. This Graphics Unit (GU) is useful at the low end of special algorithms for pixel manipulation. Geometric transformations and other manipulations are very well covered by the high–speed floating point and integer unit, which can be accessed simultaneously. Additionally the floating point adder and multiplier can operate in parallel, although this requires hand–optimised assembly routines. This goes as well for routines which want to use the graphics unit effectively. See figure 1.

All compilers use the GU only to transfer data from floating point registers to integer registers and vice versa. So the usage of the GU in shading algorithms requires always that at least part of the program code is assembly code.

The I80860 supports pipelining with special instructions and may therefore be considered as a vector processor. Again the compilers do not use pipelining (with the exception of the PSR vectoriser) and the user who wants to use pipelined operations has to manage the different pipelines and keep track of their actual status himself. This is not trivial because the I80860 has 5 different pipelines with different numbers of stages:

a) Single and Double precision FP–Adder Pipeline with 3 stages
b) Single precision FP–Multiplier Pipeline with 3 stages
c) Double precision FP–Multiplier Pipeline with 2 stages
d) Graphics Unit Pipeline with 1 stage
e) FP–Loader Pipeline with 3 stages.

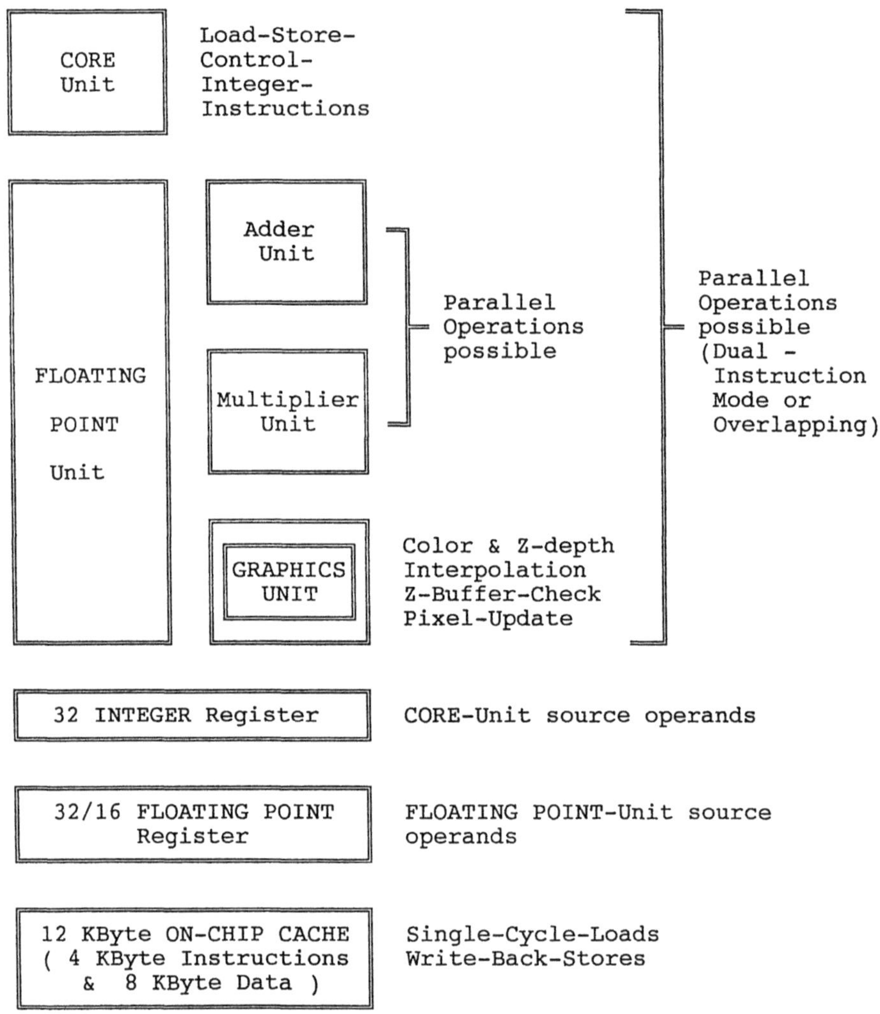

Figure 1: Hardware Layout of the INTEL I80860

Operating the I80860 in dual instruction mode, where a core unit instruction (integer arithmetic or load–store instructions) can be executed parallel to a floating point instruction (FP–Add or FP–Multiply or GU–instruction) is not supported by the standard compilers. This is not always a big disadvantage because the I80860 handles overlapping instructions itself, e.g. a floating point instruction which takes 3 cycles for completion maybe followed by 2 core instructions without additional time penalties, i.e. all three instructions will have been executed after the 3 clocks. For the assembly programmer it is nevertheless save and easy to use the dual instruction mode if he desires, especially if the pipelined versions of the floating point instructions are used. These are effectively single clock instructions and no overlap–effect can occur. The only way to achieve parallelism between core and floating point instructions is then the dual instruction mode.

When used as a vector processor, the I80860 needs (as all vector processors) special preparation of data, like gathering a large amount of input data before processing them in a program using pipelined instructions, or scattering these data after the processing etc.. With its comparatively small data cache (8KByte) and its medium sized register set (32 integer and 30 single or 15 double precision floating point registers) the I80860 is by design not a load–store machine. The investigations were performed on a KONTRON SCB860 AT–Add–On board. When this investigations were made, the on–board memory could not be operated at full speed for several reasons (only A2–steppings of the processor were available, no NENE = NExtNEar signal was generated, not the fastest RAMs were available etc.). So the data preparation turned out to be quite costly (in terms of execution speed) in some cases and the time used for this data preparation was included in the following execution speed investigations. One should keep that in mind before discarding vectorised algorithms which may not be as satisfactorily as expected (or hoped). Additionally it is not guaranteed that the presented algorithms are the fastest possibilities, especially if they involve a lot of different memory accesses. Sometimes regrouping of memory accesses is of major impact on the execution time and "programming by the book" (e.g. the "Programmer's Reference Manual"[1]) is less important than minimisation of page faults etc.. Nevertheless I have not always bothered to try every combination as long as the performance was not too unsatisfactorily.

2. Description of the Basic Algorithm

From the nature of the GU instructions as described below, it becomes clear that the supported algorithms are scanline–based algorithms for generating smooth–shaded (Gouraud–shaded) polygons. When confronted with such a powerful floating point machine one might wish to use more sophisticated rendering algorithms, like ray–tracing, radiosity algorithms etc.. It should not be excluded, that the GU may be useful for such algorithms as well, but this is far from obvious. The aim of this article is to investigate the capabilities of the GU and I will therefore restrict myself to the algorithm for which the GU instructions are obviously tailor-made.

The examinations only consider the basic interpolation, Z-Buffer and pixel-update mechanism which is typical for pixel oriented shading algorithms (e.g. scanline-algorithms). For a certain amount of pixels, the color and z-depth values are calculated by linear interpolation. After that the z-depth value for each pixel is compared with the corresponding Z-Buffer value and if necessary, the color and Z-Buffer values for this pixel are updated. The algorithm will be referred to as painting or rendering algorithm.

3. GU Instructions and Their Data Formats

The GU of the I80860 supports several different data formats. In general the color and z-depth values are "floating-point" like in the sense that they consist of an integer and a fractional part "iiii.ffff" (fixed-point format). Depending on the size one or more of the values are stored in a 64-Bit word. After the interpolation they are truncated to their integer part and packed into a special "MERGE" register from where they can be transferred to an ordinary FP-register and then stored or compared to similar values (Z-Buffer check). The supported formats are:

("fs1", "fs2" and "freg" stand for any double precision FP-register and "(p)" indicates the modified instruction for the pipelined version, if available)

3.1 8-Bit Pixel

The interpolation values are 16-Bit deep with an 8-Bit integer part and an 8-Bit fractional part. The interpolation instructions are:
"(p)faddp fs1,fs2,freg".

3.2 16-Bit Pixel

The interpolation values are 16-Bit deep with a 6-Bit integer and a 10-Bit fractional part. A 16-Bit pixel is understood to be a RGB-value with a 6-Bit R-part, a 6-Bit G-part and a 4-Bit B-part packed into a 16-Bit word. The instructions are the same as for the 8-Bit pixel, but will behave differently during the merging process. (The instructions consult a bit field in one of the control registers, where the pixel size has to be specified.)

3.3 32–Bit Pixel

Again the pixel value is an RGB–value, where each part covers 8 bit and the most significant byte is void. The interpolation value for each color is 32–Bit deep with a 8–Bit integer portion and a 24–Bit fractional part. Again the instruction is the same and the merging will be performed appropriately.

3.4 16–Bit Z–Buffer

The instruction

"(p)faddz fsl,fs2,freg"

performs a 32–Bit z–depth interpolation, where the results are truncated to their 16–Bit integer part. The

"(p)fzcks fsl,fs2,freg"

instruction performs the corresponding Z–Buffer check. This instruction stores a 64–Bit word which contains the four smaller z–values. Additionally it sets a Pixel–mask bit field in a control register which tells a consecutive pixel store instruction

"pst.d fsl,target–address"

which pixels should be updated and which not.

3.5 32–Bit Z–Buffer

If 32–Bit interpolation is required, the usual integer adds can be used. For 64–Bit interpolation (32–Bit integer and 32–Bit fraction) the

"(p)fiadd.dd fsl,fs2,freg"

instruction is available. The corresponding Z–Buffer check instruction is

"(p)fzckl fsl,fs2,freg".

3.6 Possible Combinations

The following combinations are possible (in principle, but not all are reasonable):

I) 8–Bit PIXEL & 16–Bit Z–BUFFER.

II) 8–Bit PIXEL & 32–Bit Z–BUFFER (32–Bit interpolation)

III) 8–Bit PIXEL & 32–Bit Z–BUFFER (64–Bit interpolation)

IV) 16–Bit PIXEL & 16–Bit Z–BUFFER.

V) 16–Bit PIXEL & 32–Bit Z–BUFFER (32–Bit interpolation)

VI) 16–Bit PIXEL & 32–Bit Z–BUFFER (64–Bit interpolation)

VII) 32–Bit PIXEL & 16–Bit Z–BUFFER.

VIII) 32–Bit PIXEL & 32–Bit Z–BUFFER (32–Bit interpolation)

IX) 32–Bit PIXEL & 32–Bit Z–BUFFER (64–Bit interpolation)

4. Realisations of the Basic Algorithms

Three different types of programs will be analysed and examined in the following context:

a) Pure C–Versions (PCV)

b) C–Frames with hand–optimised Core–routines (OCV)

c) C–Frames with hand–optimised Graphics–Unit–routines (OGV)

Before the presentation of the results, some general remarks on the level of optimisation and the complexity of the programs are necessary.

The accuracy of the a) and b) versions is much higher, because all interpolations are done in double precision floating point format. It is questionable whether this is really an advantage or only an assurance, but it is mentioned for the sake of honesty.

4.1 PCV

The C–routines are in general very simple and leave not much space for optimisation or variation. The only decision one has to make is whether to do the color interpolation always or only if the Z–buffer check yields a positive result.

E.g. USE :

```
for(i=0;i<no;i++) {
  z += z_delta;
  col += col_delta;
   /* Color interpolation */
  if( (zs = (short)z) < zbuf ) {
    zbuf[i] = zs;
    pixel[i] = (char) col;
  }
}
```

OR USE:

```
for(i=0;i<no;i++) {
  z += z_delta;
  if( (zs = (short)z) < zbuf ) {
    m = (double) i;
    zbuf[i] = zs;
    pixel[i] = (char)(col_start + m*col_delta);
  }
}
```

The first version will always do one floating–point add, whereas the second version will do instead only an occasional integer–to–double conversion in connection with an floating-point add and multiply. For the 8-Bit-per-Pixel cases both versions are equally fast. If more than one color value has to be interpolated, as for the RGB-cases (16- and 32-Bit-per-Pixel), the second version becomes more effective and was used in the comparisons.

4.2 OCV

In this versions the C-frame program is the same as for the PCVs, only the core-program, e.g. the "for"-loop described above, is substituted by a hand-optimised assembly routine. These routines are not too complex and any programmer with a bit of experience will not encounter greater difficulties. Again the routine for the RGB-cases is more complicated, because in this cases it pays to use pipelined floating point instructions and consequently dual instruction mode as well.

4.3 OGV

In order to use the graphics unit effectively, a lot of more or less complicated preparations are necessary. For all cases the pixels have to be treated in 8-Pixel packages to ensure that the Z-buffer check instructions and the pixel stores are synchronous.

4.3.1 Preparation of the Starting Values

All starting values have to be prepared and converted into the data formats described above. This is trivial and not time critical because it applies only for the starting and delta values.

Example:

```
/* 8-Bit-Pixel color starting value   */
/* assuming all 8 pixel use the same delta
increment */
/* "shift" double value by 8 Bit and truncate
to short */
col = (short) (col_start * 256.0);
cold = (short) (col_delta * 256.0);
```

```
/* Accumulate 8 pixels in two doubles */
double col_array[2];
cpt = (short *) col_array;
/* first the even pixels */
cpt[0] = col - 8*cold;
cpt[1] = col - 6*cold;
cpt[2] = col - 4*cold;
cpt[3] = col - 2*cold;
/* then the odd pixels */
cpt[4] = col - 7*cold;
cpt[5] = col - 5*cold;
cpt[6] = col - 4*cold;
cpt[7] = col - 1*cold;
/* Accumulate the corresponding delta values */
double cold_array;
cdpt = (short *)&cold_array;
/* Even and odd pixels use the same delta */
cdpt[0] = 8*cold;
cdpt[1] = 8*cold;
cdpt[2] = 8*cold;
cdpt[3] = 8*cold;
```

Similar preparations have to be done for the z-depth interpolation or in cases with different pixel depths.

4.3.2 Preparation of the Delta Switches

Under realistic assumptions it is not likely that the interpolation uses a constant delta value for more than 10 pixels (typically the distance from one edge to the next). The optimised routines are only efficient if much more pixels can be treated successively, at least a few 8-pixel blocks are required. Interpolating the pixels across a whole object surface would meet this requirement more likely, but to guarantee the correct interpolation it must be possible to switch from one delta value to the next. During my investigations this problem turned out to be very crucial because without a very efficient solution to this problem, the most efficient OGV routine might be completely worthless and even slower than the PCV. The next chapter will give a more detailed description of this problem and offer some solutions as well.

In general the data preparation and the algorithm depend on the chosen model (cases I) – IX)) but the overall structure is always the same :

1st Step:

Do all interpolations for color and z-depth pipelined if possible and do all required loads and stores in parallel. The special interpolation instructions merge the results as described in [1]. For 8-Bit pixels one has to interpolate alternating between even and odd pixel-numbers, always 4 Pixels per instruction. For 16-Bit and 32-Bit pixels, one interpolates 4 successive pixels per instruction, but alternating between R-, G- and B-interpolation.

2nd Step:

Do the required Z–Buffer checks for all 8 Pixels to prepare the Pixel–Mask for the subsequent stores.

3rd Step:

Do the pixel stores which interpret and reset the Pixel–Mask. This step may be executed parallel to Step 1, for overlapping loops. In this case the stores would complete the calculations of the PREVIOUS iteration and must be performed before the new Z–Buffer checks affect the Pixel–Mask setting. For steps 2 and 3 see figure 2.

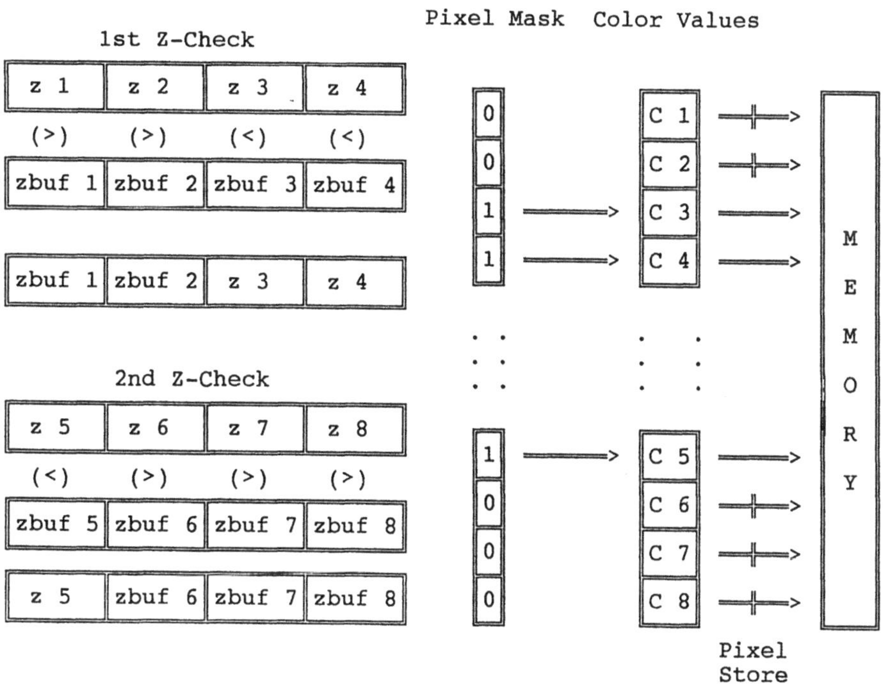

Figure 2: Z-Buffer-Check and Pixel-Store

5. The Delta–Switch Algorithm

The basic assumption for all the following considerations is that the calling program supplies all different delta values which occur along the intersection of the scanline with the object surface. This is a reasonable assumption and means no additional effort compared to the simple C–program versions, even if the data preparation is a bit more complex.

With every iteration, the OGV–Program will now load a certain delta value, use it for interpolation and then load the next value. This may be the same as the first one or a different new delta value, if at this point a delta switch is required. To avoid duplication of data, one could realise this behaviour by gathering the data into a delta structure (this should happen during the preparational period) and supplying a pointer array to the OGV–routine.

In this array a pointer to a certain delta structure appears as often as this structure is needed for interpolation. A delta switch is then nothing else than changing from one pointer to the next. In reality this procedure is not easily implemented for the following reason: All delta–values have to be gathered in double (64–Bit) words, because all pixel manipulation instructions operate on double words.

In detail this means:

Four 16–Bit Z–depth or color values, or two 32–Bit Z–depth or color values, or eight 8–Bit color values must be accumulated in a double word. Depending on the combination of Z–Buffer and color values, such a delta structure would contain up to 8 different delta values. Taken into account that one delta value might be used only for about 10 Pixels, it becomes clear, that in this case the preparation of the delta structures does not lead to data multiplication. Consequently, for each different model one has to find the best suited solution, which in general is a mixture of using extended delta–value arrays and delta–structure–pointer arrays. (The extended delta–value array maybe a structure array as well, so the emphasis is put on the term "pointer" in the second expression.)

Timing these models may be difficult, because it is not always obvious, which preparations are really additional efforts and must be included in the execution time measurements. So for the three examples which are considered below, the algorithm which is used is explained in detail. Further clarification is given which parts of the algorithm are included in the execution time measurements. It is common to all applications using "vectorised" special purpose routines, that the data preparations should be subject to careful considerations, because otherwise they might cost more than the optimised routines gain.

5.1 Examples

With 16–Bit pixels and 16–Bit z–Buffer the smallest complete delta structure would consist of:

```
2 double words, each containing two z-deltas
3 double words, 1 for each color with 4 deltas per double word
-------------------------------------------------------------
10 words per structure.
```

If the average pixel chain length is 200 pixels with a delta switch every 10th pixel, this would require a total of 30 different delta structures. For two successive sets of pixels one would need two "pure" structures which will be used twice and one "transition" structure containing the delta switch, i.e. a total of 20 pure structures and 10 transition structures. Compared to the original set of different delta values (20 x 2.5 words) the new set requires 6 times the memory space and the corresponding load–store operations.

The pointer–array method would need this set of different structures and additionally a pointer array, where each pure structure is referenced twice and each transition structure only once. This would mean another 50 words of memory space and the corresponding store operations. An asynchronous treatment of z–delta and color–delta values would be another possibility one could think of, but for the model with 32–bit z–interpolation this is disadvantageous :

```
30 different color-delta structures, 6 words each = 180 words
50 color-delta-pointers,             1 word each  =  50 words
20 different z-delta values,         2 words each =  40 words
100 z-delta-pointers,                1 word each  = 100 words
-------------------------------------------------------------
TOTAL                                             = 370 words
```

compared to 350 words for the simple pointer array.

For the extended delta–value array method the pure structures would have to be duplicated to yield the complete set of 50 delta structures. The additional effort would amount to 200 words memory space accompanied by the corresponding load–store operations. In this example the pointer method should prove to be superior, as it is probably for all cases, where the pixel size is larger than 8 Bit. Only if the pixel subsets along which a delta value remains constant become very small (e.g. 4 pixels or less) the delta–value array method will be advantageous. In this case the set of different delta structures will already be the extended delta–value array.

5.2 Summary

To enable efficient usage of optimised pipelined pixel operations it is necessary to maintain a continuous input flow of delta values for the interpolations. If the usage of extended delta–value arrays does not involve to many preparational load–store operations it is a problem independent solution which is easy to include in the execution time measurement.

The usage of pointer–arrays will involve some additional load–store operations as well, but this effort is difficult to estimate. If the number of delta switches is large the method is not very efficient, especially if the values for more than 2 pixels must be packed into double words.

6. Results

Three of the 9 different cases mentioned above (cf. 3.6) have been investigated for the following reasons:

a) Case I:

This case is the classical example for using the delta–value method of implementing the delta–switch algorithm.

b) Case IX:

This case is quite the opposite to Case I) and is best suited for the pointer–array approach. The time measurements are based on the simplified model described in chapter 5.1 (Example).

c) Case V:

Because of the simple z–interpolation this case yields the fastest OCV programs and offers therefore the biggest challenge for the Graphics Unit.

For the OCV and the PCV programs the execution time depends on the size of the screen and the number of substitutions due to positive Z–Buffer compares. The latter is also a function of the screen size, if the screen size is small, the Z–Buffer becomes loaded with small values quite soon and the number of substitutions decreases. The screen size for PCV and OCV programs was 768 x 1024 pixels. The execution time of the OGV is independent of the number of substitutions, because the store instructions will always be executed, but realised only where the Pixel–Mask–Bits are set.

The execution time measurements presented below, are given for various pixel chain lengths, from 10 to 1000. Only for the OGV programs pixel chain lengths larger than 10 may model real situations. For OCV and PCV programs the corresponding figures are given only for the sake of completeness.

Each run computes 100,000 different pixel chains of the given length, with the Z–depth values randomly distributed. To enable a correct comparison it must be noted that e.g., 100,000 repetions of pixel chain length 200 for the OGV are equivalent to 2,000,000 repetitions of pixel chain length 10 for the PCV and OCV.

The EQUIVALENCE COMPARISON after each table gives the corresponding figures. The number of substitutions for the OCV and PCV programs is app. 2,200,000 in this case.

6.1 Case I)

No of repetitions : 100,000

Table 1

No of Pixels	OGV	OCV	PCV
10	----	940	1820
20	1260	1600	3130
40	1430	2850	5490
100	3900	6430	11860
200	6540	12350	21810
1000	40100	58710	99420

All timings are given in Milli-seconds.
EQUIVALENCE COMPARISON:
OGV : 100,000 x 200 in 6540 msec
is equivalent to
OCV : 2,000,000 x 10 in 15680 msec
PCV : 2,000,000 x 10 in 25350 msec
with 2,200,000 substitutions.

In terms of Gouraud-shaded polygons, the latter results may be translated into:
OGV : 30,500
OCV : 12,750
PCV : 7,900
100-Pixel GOURAUD-SHADED POLYGONS per SECOND.

6.1.1 Remarks

The delta-switch algorithm is based solely on delta-value arrays, because in this case each delta-structure would contain the delta values for 8 successive pixels gathered in a total of 4 double words.

Optimised copy routines are used to distribute and repeat the delta values appropriately. The time needed for this copies is INCLUDED in the execution time measurements. Not included is the preparation of the index array, which contains the number of pixels for which each delta value is valid. This array is used by the optimised copy routines. (The array contains nothing else than the number of pixels between an edge pair and is therefore assumed to be supplied by the calling program, which will store these data anyway.)

The original values are assumed to be supplied in two different arrays, one containing the 32-Bit z-delta values (format: 16-Bit integer and 16-Bit fractional part) the other containing the 16-Bit color values (format: 8-Bit integer and 8-Bit fractional part). Two different copy routines are required to generate the extended arrays.

The OCV program uses only simple optimisation, because of the small amount of floating point calculations. Pipelining and dual instruction mode are no improvement in this case and the optimisation is achieved by ordering and minimising the number of instructions.

6.2 Case IX)

No of repetitions : 100,000

Table 2

No of Pixels	OGV	OCV	PCV
10	. ----	990	2810
20	1320	1590	4610
40	1760	2800	7470
100	4220	6150	13790
200	8070	11480	22680
1000	61130	53160	86340

All timings are given in Milli-seconds.

EQUIVALENCE COMPARISON:

OGV : 100,000 x 200 in 8070 msec

is equivalent to

OCV : 2,000,000 x 10 in 15640 msec

PCV : 2,000,000 x 10 in 25260 msec

with 2,200,000 substitutions.

In terms of Gouraud-shaded polygons, the latter results may be translated into:

OGV : 24,800

OCV : 12,750

PCV : 7,900

100-Pixel GOURAUD-SHADED POLYGONS per SECOND.

6.2.1 Remarks

In this case all delta values are gathered into one large delta structure of the type: "struct delta { double z1-delta, z2-delta, R-delta, G-delta, B-delta };"

This delta-structure array is twice as large as the original, because two color values have to be accumulated in one double word. It is clear that this method simplifies addressing, because all delta values are close together and some time will be saved, due to cached loads in the OGV-Program.

The execution time measurement INCLUDES the time needed to produce the delta structures from the original data.

The example of chapter 5.1 is of course only a very crude model for reality. One of its artificial properties is that it needs no transition delta structures, but even in real situations there should be not too many of these structures. Provided the program permits an easy way to insert such transition structures (e.g., by adding a pointer "struct delta *next" to the "struct delta" so that the different structures are forming a chain rather than an array), this means not much additional effort. One could use the optimised routine to produce the pure delta structures by duplication and afterwards insert the transition deltas at the appropriate locations.

Additional investigations have shown that an asynchronous treatment of the z-delta and color-delta values is disadvantageous, it needs more memory space and is slower.

The PCV program is slower as in case I) as one could expect because of the additional effort of interpolating three color values. In case of the EQUIVALENCE COMPARISON this difference becomes neglectable due to the large number of memory accesses.

The OCV program is as fast as in case I), because the larger amount of floating point operations allows the efficient usage of pipelined instructions in connection with dual instruction mode.

The OCV becomes than more or less independent of the number of substitutions, because the almost all calculation are finished anyway before the result of the Z-Buffer check affects the color interpolation. (The Z-Depth value has to pass through the Adder-Pipeline twice, before the Z-Buffer check can be performed.)

6.3 Case V)

No of repetitions : 100,000

Table 3

No of Pixels	OGV	OCV	PCV
10	----	880	2580
20	1810	1380	4230
40	1490	2250	6420
100	4120	4510	10880
200	5990	7960	16360
1000	63770	34270	52220

All timings are given in Milli-seconds.

EQUIVALENCE COMPARISON:

OGV : 100,000 x 200 in 5990 msec

is equivalent to

OCV : 2,000,000 x 10 in 11860 msec

PCV : 2,000,000 x 10 in 18200 msec

with 2,200,000 substitutions.

In terms of Gouraud-shaded polygons, the latter results may be translated into:

OGV : 33,300

OCV : 16,850

PCV : 11,000

100-Pixel GOURAUD-SHADED POLYGONS per SECOND.

6.3.1 Remarks

With 2 z-delta values or 4 color values per double word the complexity of this model lies between the previous cases.

The 32-Bit Z-Buffer with 32-Bit interpolation means that simple integer adds can be used for the z-depth interpolation. This has its effect mainly on the OCV-program, were z-interpolation can now be performed parallel to the color interpolation and no float-to-integer conversion for the z-values is required. This explains the extraordinarily good OCV performance. More or less the same goes for the PCV program.

The OGV program behaves a bit irregular. The drops in execution time for pixel chain length 40 and 200 are due to the fact that these numbers are multiples of 8. This means that all pixels are updated with the optimised algorithm. If the pixel chain length is less than 16 and no multiple of 8, all or at least the modulo-8-rest pixels are treated as in the PCV program. The amount of optimisation for the OGV program must be very large to achieve a better performance as the OCV. Again the method of the delta-structure-pointer array proved to be most efficient.

I have tried several other possibilities as well, e.g. separate z-depth interpolation, delta-value arrays etc., but the pointer array method gave the best result, if a high level of optimisation is provided. Not only the pixel update routine has to be optimised, but the data preparation routines as well.

The execution time measurement INCLUDES the complete preparation of the delta structures, which in this case means transition structures as well, and the generation of the pointer array. The underlying model is exactly the model described in chapter 5.1, which means that it is still a simplification. For a real model the optimisation of the data preparation routines becomes a bit more complex, but the strategy may be the same:

1. step :

Generate all pure delta structures. This is simple and easy to optimise. The delta structures should be connected via pointers and may include the number of occurrences (i.e. how often this structure is to be used for interpolation).

2. step :

Insert all transition structures. For the simple model this is schematic as well and therefore easy to optimise. For real situations it may become tedious, but nonetheless the number of transition structures should remain small. Therefore it consumes not to much execution time, even if it is programmed in C-code.

3. step :

Generate the pointer array to be used by the pixel update routine. This can be optimised using different strategies, each based on the occurrency information contained in the delta structures.

The OCV and PCV routines profit from the very simple Z-Buffer interpolation. Especially for the OCV routine which is in principle the same as for case IX) the benefit is large. The z-depth interpolation is now a Core operation and can be performed parallel to a floating point operation. The Z-Buffer check can therefore be performed much earlier and consequently for negative compares aborts the color interpolation much earlier.

7. Miscellaneous

The rendering algorithm is only one example, where the Graphics Unit instructions are useful. One may think of other examples, where the interpolation features are useful as well, e.g. in line drawing algorithms.

But one should keep always in mind, that the conversion of the input data (which may be in "float" or "double") to the special format needed for the interpolation instructions requires additional effort. It is therefore only efficient if the subsequent calculations make extensive use of the converted data.

In general preparation of a single data item requires

1 floating point load of the input values and

1 floating point multiply,

in order to "shift" the input values and save the fractional part after

1 truncation to an integer format

and 1 integer store is performed.

(Almost always some additional manipulations are necessary, because the interpolation instructions of the GU do more than one interpolation with each instruction. This means one has use larger increments, more than one starting value, etc..)

Example:

16 Bit Z–Buffer value:

```
double input;
long output;
/* Shift by 16 Bit before conversion */
output = (long) (input * 65536.0);
```

In optimised code the conversion takes between 4 and 7 clock cycles, depending whether the load and store hit the data cache or not. For the application this means, that any program not using the GU may take 4 to 7 clock cycles more time. If one uses highly optimised assembly code this could be equivalent to 4 to 7 operations in dual instruction mode. For a simple line drawing algorithm, i.e. the interpolation between a starting point and an end point on the screen, a optimised routine using the GU is not necessarily faster than an optimised routine not using the GU, but the latter requires less programming effort.

In this case the scale could tip towards the GU program, if the lines to be drawn are quite long (e.g. in the range of 100 Pixels) or if additional use of the data would involve compares and corresponding pixel updates, etc.. One might think of a clever way to implement line styles this way or something like that.

This issue will not be pursued any farther in this paper, it is just an example for the general rule:

The usage of the Graphics Unit's interpolation and merging instructions is only reasonable, if the calculation effort justifies the data conversion.

8. Conclusions

With a considerable amount of optimisation effort it is always possible to accelerate the painting algorithm using the GU instructions in a special assembly routine.

In general this goes hand in hand with optimised data preparation routines and results in complex algorithms. The acceleration factors lie between 2 and 3 for OCV and between 3 and 4 for PCV routines.

The execution time does not depend much on the complexity of the underlying models, for the examined cases it was more or less in the same range. E. g. the choice between 8–Bit pixels and 16–Bit or 32–Bit pixels affects the complexity but not the performance of the resulting program.

The OCV method offers an efficient and comparatively simple possibility for optimisation, if the problem is not too time critical or if development time is more critical than performance. Especially for the simple Z–Buffer interpolation it is a good alternative.

When operated as a graphics engine, the GU of the Intel I80860 is only of limited use, ie. for 2D– and 3D–Output primitives of the kind described above. But it may play a decisive role if the principle "Simple Things should be Fast" is to be fulfilled, even so it requires a lot of optimisation.

9. References

1. i860 64-Bit Microprocessor Programmer's Reference Manual. Intel Literature Sales, Santa Clara, CA 95052-8130,1989

A Chinese-Character and Graphics Workstation

Shi Jiaoying, Huang Jianfeng, Liu Liancai and Hu Jingyi

Department of Computer Science & Engineering.
Zheijang University
China

Abstract

This paper introduces the design approaches of a Chinese-character and graphics workstation DGS-8000 which has been developed at the Computer Graphics Research Laboratory, Department of Computer Science and Engineering, Zhejiang University. The specifications, architecture, Chinese-character processing environment and working modes of the workstation are described. Especially the I/O interface with the host computer, the processor unit, the graphics generation and display unit, Chinese-character operating system and some main input modes of Chinese-character codes are introduced in detail.

1. Introduction

Due to very rapid development of IC and computer industries, a new type configuration of computer system, namely graphics workstation, has been appeared since early 1980s. The graphics workstation can be characterized as a computer system with sufficient computing power, graphics capabilities and networking ability. Now the graphics workstations are getting popular and popular in engineering application fields, so sometimes they are called as "engineering workstations". The graphics workstations can be used alone to solve engineering problem with certain scale, but also they can be used as a node in computer networks or as a graphics terminal of a host computer very flexible.

This article introduces the design approaches of a high resolution graphics workstation DGS-8000 with Chinese-character processing capabilities, which has been developed at the Computer Graphics Research Laboratory, Department of Computer Science and Engineering, Zhejiang University. The works are based on the previous experiences of the Lab [1], [2], [3], [4]. It is a low-cost high performance graphics workstaton, which will be used primarily as a high performance graphics terminal and a intelligent terminal with Chinese-character processing capabilities of the VAX-11/785 in our Lab. But as a homemade Chinese-character and graphics workstation, we think that in the next step the workstation should be commercialized to satisfy the domestic needs.

2. Specifications of the DGS-8000

The DGS–8000 workstation is compatible with the most popular personal computer IBM–PC/XT and runs under Microsoft DOS operating system and own monitor for use of interactive graphics. The workstation can be configured with various peripheral devices such as digital tablet, plotter, mouse and printer. The DGS–8000 may be interfaced with the host computer via two ways, i. e. standard RS–232C serial port and DMA channel with VAX–11 or PDP–11 series machines. The typical specifications of DGS–8000 are:

Graphics Resolution:	1024×1024 (1024×768 visible) .
Color Resolution:	8 bits, 256 colors displayed simultaneously.
Color Palette:	16. 7 million colors.
Overlay Bitplane:	1024×1024×1 bit.
Scan Mode:	Interlaced or Noninterlaced.
Monitor Format:	19–inch Color.
CPU:	8088, 8 MHZ.
Memory:	768 KB.
Floppy Disks:	2×360 KB.
Hard Disk:	20 MB.
I/O Interface:	RS–232C and DMA.
Operating System:	DGS–DOS Compatible with Microsoft DOS.
Chinese–Character Operating Systems:	DGS–CCDOS and DGS–LXCCDOS.
Chinese–Character Font Library:	Hard font library includes the National Standard Chinese–character Font Libraries of the I and II classes, which contain 8192 Chinese–characters and 616 graphics symbols.
Input Modes of Chinese–Characters:	4 modes under DGS–CCDOS and 12 modes under DGS–LXCCDOS.

3. Architecture of the DGS-8000

The DGS–8000 workstation is consist of three units, i. e. (1) I/O interface with the host computer; (2) Processor unit; (3) Graphics generation and display unit. The logic block diagram of the DGS–8000 is shown in Fig 1.

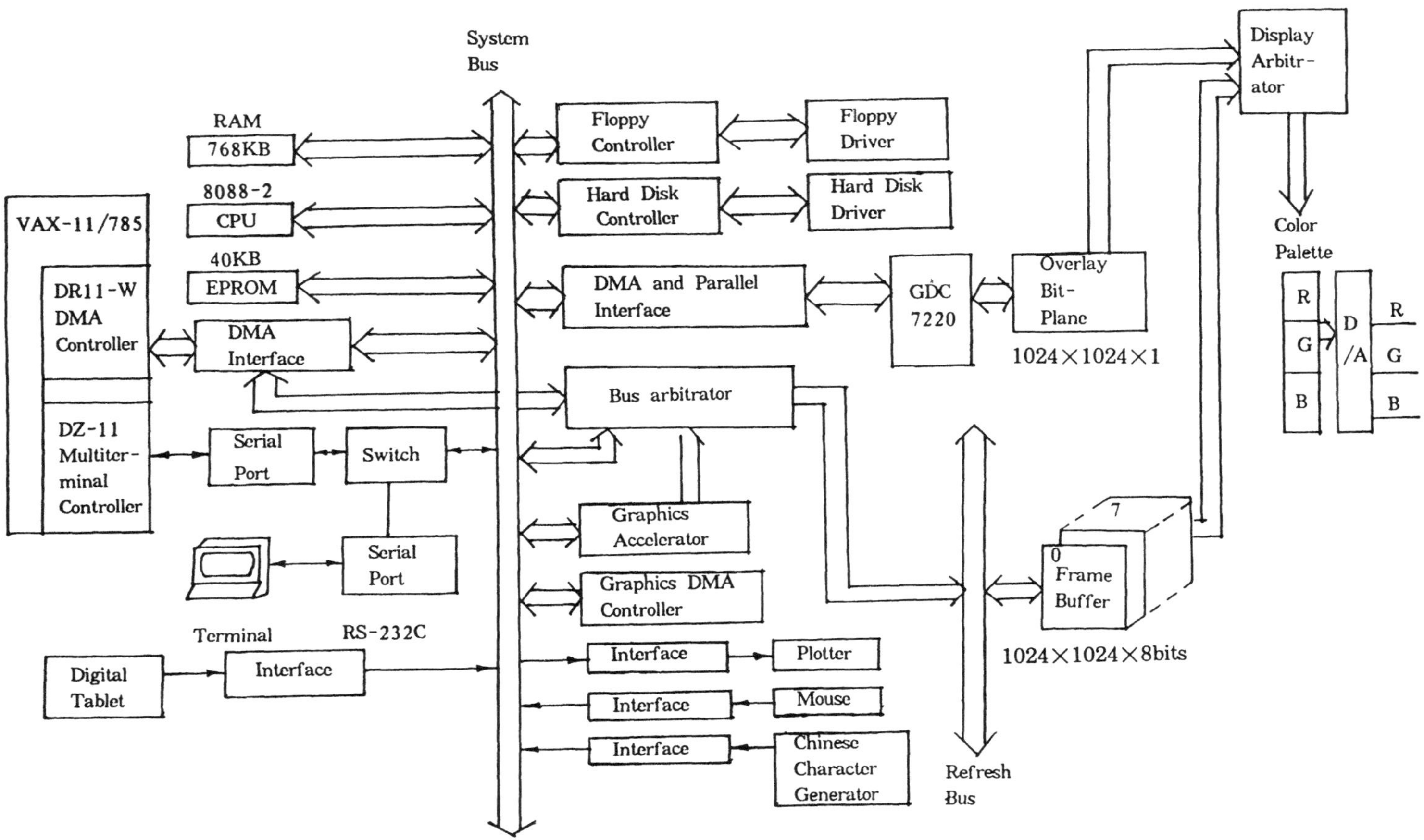

Fig. 1 DGS-8000 block diagram

3.1 I/O Interface

The DGS-8000 provides two I/O interface modes, the serial port and DMA channel, as mentioned in the section 2. The serial port RS-232C is a standard interface which allows DGS-8000 to communicate with any kind of computers, but the data transfer is too slow via serial port. In order to increase the speed of data transfer between the host and the workstation, there is a DMA channel which can be used to deliver graphics commands, as well as to transfer the graphics data which have been generated by the host and stored in the main memory of the host to the frame buffer of the workstation or vise verse.

The workstation provides various working modes under control of the console commands. The operator needs only to input on-line or off-line commands on the keyboards, the workstation can be switched flexibly from one working mode to another.

3.2 Processor Unit

The processor unit of the DGS-8000 may be divided by two parts, the core part and the expanded part. The core part of the processor unit is actually a IBM-PC/XT compatible processor. The INTEL 8088 microprocessor, the IBM-PC 62-pins bus standard and PC-DOS are adopted. All these features ensure fully compatibility of our workstation with IBM-PC's existing software and hardware. All PC's peripheral cards can be plugged in our workstation, and all software packages run in IBM-PCs will be run in DGS-8000 perfectly.

The Fig. 1 shows that DGS-8000 provides various peripheral devices and special data transfer channels. Therefore one piece of interrupt controller chip (8259) and one piece of DMA controller (8237), original PC design scheme adopted, are dissatisfied in our design. DGS-8000 uses two pieces of interrupt controller 8259 in cascade modes. Except original one piece of DMA controller 8237, other two expanded pieces of DMA controller 8237 are used in our design. One expanded 8237 is used as the slave of the original DMA controller. The channels of the slave DMA controller is distributed to satisfy fast data transfer between 8088 and other parts. The four channels of the slave 8237 are as follows:

Channel 0: System RAM Refresh.
Channel 1: Command and Data Transfer from Host Computer to 8088.
Channel 2: Data Transfer from 8088 to Host.
Channel 3: Data Exchange Between 7220 and 8088.

The other expanded DMA controller chip is dedicated for communication between the host and the frame buffer in DGS-8000. The four channels of this 8237 are distributed as follows:

Channel 0 and Channel 1: They are used for data block transfer with in frame buffer. It results that the contents of one area on the monitor's screen can be transferred very quickly to another area on the screen.

Channel 2: The host reads the data from frame buffer.

Channel 3: The host writes the data to frame buffer.

In the expanded part of the processor unit we added a piece of special hardware for Chinese-character processing which includes two VLSI chips of Chinese-character base and its controller and associate logic. The Chinese-character are displayed on the overlay bit-plane which has highest display priority, so the mixed effect of Chinese-character and image on the screen can be obtained (see Fig. 2) . It is very unique character of DGS-8000 workstation.

We have to mention that we have rewritten the BIOS of the PC-DOS to meet the requirements of expanded hardwares and Chinese-character text editor, but the new version of BIOS is still compatible with the original.

Fig. 2 Mixed Chinese-character and image on the screen

3.3 Graphics Generation and Display Unit

This unit consists of image memory, bus arbitrator, graphics accelerator, display arbitrator, color palette and D/A converter (see Fig. 1) . The image memory is the heart of this unit. It consists of frame buffer and overlay bit-plane. The frame buffer is organized as a 1024×1024×8 bits memory which is used to store image and can be read/written by 8088 cpu, the host computer and the graphics accelerator chosen by the bus arbitrator. The overlay is a 1024×1024×1 bit-plane which is controlled by a dedicated graphics display controller chip-upd 7220. The overlay is used to store wireframe graphics, cursors, alphanumerical characters and Chinese-characters. The overlay has the display priority higher than that of the frame buffer. There are three display modes, i. e. overlay only, frame buffer only and mixed mode of the both. In the mixed mode the overlay image is always in the front of the image of the frame buffer. The display arbitrator controls these three display modes.

Our memory organization of the frame buffer is different from the common bit-plane organization. In bit-plane organization each bit of the bit-plane corresponds with a pixel [5] . In DGS-8000 each pixel contains 8 bits, if bit-plane scheme is adopted, the 8 bits of the pixel will be located on 8 different bit-planes. Therefore to modify a pixel value, each bit-plane has to be read, modified and written. Obviously it slows down the operating speed. The block diagram of frame buffer organization of the DGS-8000 is shown in Fig. 3. One byte of the bit-plane corresponds with a pixel on the screen. Consequently to modify one pixel value only one bit-plane needs to be read, modified and written. The operating speed is increased.

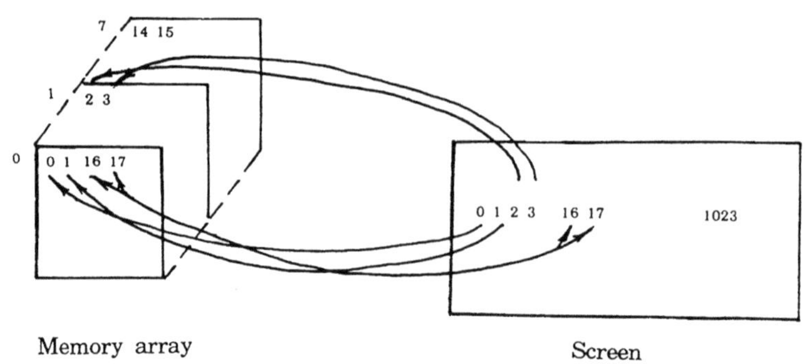

Fig. 3 Memory organization of the frame buffer in DGS-8000

The graphics accelerator is implemented using 16-bit bit-slice microprocessor 29116, timing controller 2910 and 16 bit high speed multiplier TDC-1010. All these high speed

ICs and their associated logic are assembled on a card which can be plugged into 62–pin PC–bus. The accelerator functions include hardware generation of the graphics primitives, such as lines, arcs, circles and curves etc. , and geometry transformations. And a molecular model generation algorithm developed in our lab is implemented in the accelerator using microprogramming technique.

4. Chinese-character Processing Environment

It is true that any computer system without Chinese–character processing capabilities can' t be got popularity in China. The DGS–8000 provides users a powerful Chinese–character processing environment. There are two Chinese–character operating system such as DGS–CCDOS and DGS–LXCCDOS. The origin of these two OS is the CCDOS which is the first Chinese–character operating system compatible with the Microsoft DOS. So the DGS–CCDOS and DGS–LXCCDOS are compatible with the DOS too. The hierarchical structure of the CCDOS is shown in Fig. 4.

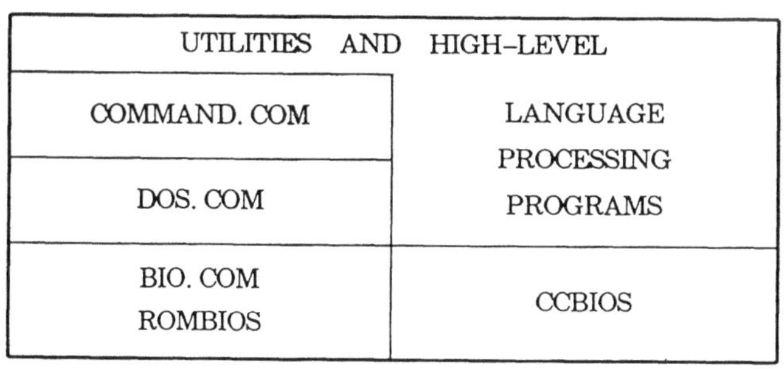

Fig. 4 The structure of the CCDOS

From Fig 4. the difference between DOS and CCDOS is the additional part of the CCBIOS which is the unique expand part of the ROMBIOS. CCBIOS includes several driver modules such as Chinese–character keyboard input, Chinese–character display and Chinese–character printing etc. . We have transplanted CCDOS to our DGS–8000 workstation hardware and software environment and a series of modifications have been introduced. The font library management, display management, keyboard management, communication management and printing modules are all rewritten completely. The structure diagram of the DGS–CCDOS Chinese–character processing module is shown on Fig 5.

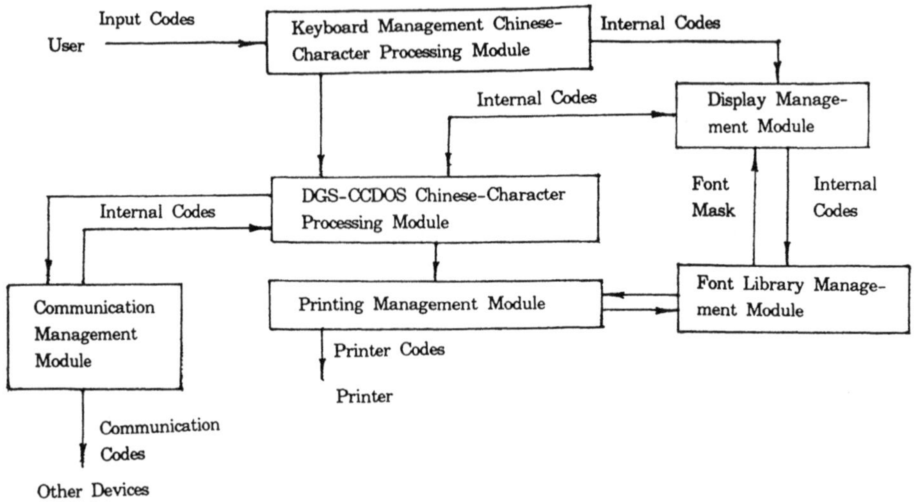

Fig. 5 The structure of the DGS-CCDOS' Chinese-character processing module

The screen display format consists of 25 rows and 40 Chinese characters per row. The hard font library is a set of two custom design chips. One is a 2 MBytes ROM which contains 8192 Chinese characters in 16×16 dot matrix form each. The another is a decode chip. The advantages of using hard font library are the high access speed of the dot matrix data, easy to program and no memory space needed.

DGS-8000 Chinese-character processing environment includes UNIX' s vi simulator when DGS-8000 communicates with the host computer via RS-232C. It means that DGS-8000 can simulate not only UNIX' s vi full screen text editor in English, but also can simulate UNIX' s vi text editor in Chinese-characters. As we know vi text editor under UNIX is a very powerful tool, so that our vi simulator in Chinese-character also is a highlight in Chinese-character processing environment.

5. Some Main Input Modes of Chinese-Character Codes

In Chinese information processing area more than four hundred kinds of Chinese character encoding modes have been developed in China now, but only a few encoding modes got to be used popularly. Many of the encoding modes are too difficulty to learn. The research topic on Chinese-character encoding is still very popular, because people would like to make the input of Chinese characters easier and faster. The most popular encoding modes of Chinese-character are "zip" code, pinyin code, fast pinyin code, telegram code, five-stroke code, eight-stroke code, head-tail code, English input code etc. The Chinese character encoding modes can be categorized basically into four classes, they are position based, phonetic alphabet based, character structure based and character radical based classes.

Two powerful Chinese-character operating systems can be run in the DGS-8000 workstation, i. e. DGS-CCDOS and DGS-LXCCDOS. They are developed from the original Chinese-character operating system CCDOS. Especially the DGS-LXCCDOS is the most powerful Chinese-character operating system which is able to output associate words when a single Chinese character is given. For example, when a Chinese character "人" (which means a person) was input, a series of Chinese words associated with this "人" will be displayed on the echo line of the screen, such as

"人民" which means "people"
"人群" which means "crowd"
"人口" which means "population"
"人才" which means "a talented person"
"人道" which means "humanity"
"人类" which means "mankind"
etc.

So the DGS-LXCCDOS is often called as associated Chinese-character operating system. All Chinese character encoding modes mentioned in the previous paragraphs are adopted by either DGS-CCDOS or DGS-LXCCDOS. Now we introduce several typical Chinese coding modes in a bit detail.

The "zip" code mode is a position based encoding mode. According the national standard GB2312-80 Information Interchange for Chinese Character Code, the Chinese characters and graphics symbols are positioned on a two-dimensional array with 87 rows and columns respectively. The characters of the font library class one are located in the range from row 1 to 55, and characters of the font library class two are located in the range from row 56 to 87. the character is coded by the row number and the column number of its position. For example, the character "啊" is located in row 16 and column 1, it is coded as "1601". The code "1601" is used as to input the character "啊". When "zip" code mode is used, four digital keys need to be struck to input one character.

The pinyin code mode is a phonetic alphabet based encoding mode. Pinyin mode used 26 English letters as Chinese phonetic alphabets. For example, the pronunciation of the Chinese character "中" (which means "center") is "zhong". In order to input "中", five keys "z-h-o-n-g" need to be struck. For decreasing the number of struck keys, the fast-pinyin input mode has been developed. In fast-pinyin mode some combined vowels and consonants are defined as a single key. Fig. 6 shows the definitions of the combined vowels and consonants. Now two strikes "a-s" are needed only to input the "中".

PINYIN	CORRESP. KEYS
zh	a
ch	i
sh	u
ao	k
ai	l
ü	v

PINYIN	CORRESP. KEYS
an	j
ang	h
en	f
eng	g
ong	s
ing	y

Fig. 6 Fast-pinyin keyboard

The head-tail code mode is a character structure based encoding mode. In this mode all Chinese-character can be constructed by the 52 character-roots, and these 52 character-roots are assigned to 26 letter keys. As an example, Fig. 7 shows several key assignments in this mode.

keys	character-root
a	小
	又
b	冖 宀
	山
c	士 土
	尸 户

keys	character-root
s	八
	女
w	厂
	犭
z	人
	卜

Fig. 7 Some key assignments in head-tail code

In the head-tail code mode a Chinese-character is always encoded by its left-upper part and right-lower part. For example, a Chinese character "从" can be encoded by its left-upper part "人", and right-lower part "人". From Fig. 7 we see that character-root "人" is corresponded with key "z", so striking key "z" twice will input the character "从".

The five-stroke code mode is a character radical based encoding mode. In this mode all Chinese character are divided into 132 radicals, which are assigned to 25 English letter keys from a to y, There is a song to help people to learn by heart the assignment of the 132 radicals to 25 letter keys. So the five-stroke code is a very fast input mode. A skilled typist may input about 270-160 Chinese-characters per minute.

Finally, it is very interesting that one of the Chinese character input modes uses English language as an input mode. It means when user input an English word, the screen echoes a Chinese word. Under this mode, for example a word "people" was input. then corresponding Chinese word"人民" appears on the screen. This input mode is very convenient for the people who knows English and is learning Chinese.

6 Working Modes of the DGS-8000

DGS-8000 provides various flexible working modes which can satisfy users' requirements. It may work alone either in DOS environment, under this mode plenty PC software may be called, or in interactive graphics mode, under this mode screen image can be modified interactively. On the other hand it can be communicated with host computer, under this mode the host resources can be used, the drawing commands and data can be transferred via DMA channel or serial port RS-232C.

There are 6 working modes as follows:

(1). DOS Environment: Under this mode PC's software can be run. Using soft interrupt INT 60H, 256 graphics routines can be called.

(2). Interactive Graphics Mode: The screen image may be modified interactively using keyboard, menus, mouse or tablet etc.

(3). Selftesting Mode: The selftesting routine is called.

(4). On-line, Nongraphics Mode: The DGS-8000 communicates with the host computer via RS-232C. The workstation's console simulates the host's terminal. The data can be exchanged under control of communication software.

(5). On-line, DMA Drawing Mode: The workstation's console simulates the host's terminal. The drawing commands may be sent from the host to DGS-8000, and the host generated image can be transferred into frame buffer and displayed on the screen; at the same time the contents of the frame buffer or the status the workstation can be read back to the host. When whole frame image is exchanged, the DGS-8000 provides the run-length encoding and decoding hardware to compress information volume and increase transfer speed.

(6). ON-line, Serial Port Drawing Mode: The workstation's console does not simulate the host's terminal. The host sends the drawing commands directly via RS-232C.

All these working modes can be switched from one to another very flexibly.

7. Conclusion

The design principles of the DGS-8000 are based on common IC chips and the popularity situation of the personal computers in China today. Therefore the DGS-8000 is a low-cost workstation and is easy to use. We think the graphics capabilities of the DGS-8000 may be matched with popularized version of the workstations abroad. The Chinese-character processing capabilities of the DGS-8000 are higher than them. The next step of our works is to commercialize DGS-8000 and to improve processor unit using 32 bit microprocessor.

Reference

1. Shi Jiaoying, Chen Maolin etc. "Some Innovative Application Design Approaches of GDC 7220 " Proceedings of IEEE ICCA' 87, pp. 612-618
2. J. Standhammer, J. Huang, L. Liu "High Performance Display System for Dynamic Image Generation" Proceedings of IEEE ICCA' 87 pp. 336-343
3. Shi Jiaoying etc. "A Low-Cost Raster-Scan Graphics and Image Processing System" Journal of Peripheral Devices (in Chinese) No. 2 1987 pp. 100-103
4. Shi Jiaoying etc. " Some Design Aspects of the Raster-Scan Graphics Displays " Proceedings of ICCC' 86 pp. 598-604
5. Mary C. Whitton " Memory Design for Raster Graphics Displays" IEEE Computer Graphics and Applications vol. 4 NO. 3 pp. 48-65, 1984

A Distributed Frame Buffer within a Window-Oriented High Performance Graphics System

Thomas Haaker and Harald Selzer
Technische Hochschule Darmstadt
FG Graphisch-Interaktive Systeme

Hans Joseph
Fraunhofer-Arbeitsgruppe für Graphische Datenverarbeitung

Wilhelminenstr. 7, D-6100 Darmstadt, FRG

Today's workstation users demand high computational performance combined with powerful graphics and a comfortable window system. Existing and forthcoming standards like GKS-3D, PHIGS/PHIGS+, X Window System, and PEX have to be supported optimally.

This paper presents the architecture of a graphics engine designed to meet the above requirements. Utilizing a distributed frame buffer pixel access with a high bandwidth is achieved. Several functions of a window management system like clipping at arbitrarily shaped window boundaries, fast copying of windows and performing Bit-Block-Transfer operations (BitBlT) are performed by hardware. Finally, a homogeneous and load-adaptive multiprocessor configuration for geometry and rendering calculation is described.

1. Introduction and Background

Computer Graphics is an area of research and industrial exploitation with a large and still increasing number of applications. The human visual perception is the most effective method to perceive a large amount of information. Therefore graphics methods for displaying data including efficient man-machine interfaces enhance the user's productivity. The photorealistic rendering of complex scenes and fast computations for the visualization of technical, physical, chemical, or medical data requires innovative workstation architectures. Speed improvements will be achieved by the design of specialized hardware to meet the requirements for the display of large amounts of scientific data within time constraints. Moreover the user asks for an environment that allows the simultaneous display of computational results of different applications. This has to be performed in parallel to the access and manipulation of data by several user processes. This task is best performed by the use of workstations. They provide a window environment as an interface to the user. To meet the requirements for concurrency and image quality with high interactivity, a substantial hardware support for window-oriented workstations is mandatory.

The low price of raster scan monitors and the rapid decrease in cost of semiconductor memories introduced the raster displays as the only important technique in the mass market of workstations. In the workstation area the raster scan displays are assisted by a frame buffer - often associated with the term 'image memory'. The frame buffer contains the image to be displayed on a pixel-by-pixel basis and refreshes the raster scan display continuously. This implies a memory architecture with two functional ports. One port is used by the rasterization process which computes all the pixels from a geometric description. The other port is controlled by a circuit which routes the color values of the stored pixels to the CRT display [14,17]. The advantage of a frame buffer is based on the fact that the screen refresh process is independent of the image generation and rasterization process, i.e. the refresh frequency for the monitor can be chosen independently of the frame buffer update rate achieved by the rasterization process. Therefore the frame buffer manages to display an arbitrary image with no limits to the complexity and the number of graphics objects.

On the other hand, a lot of pixels have to be updated in order to obtain major image modifications. As a consequence of today's very high resolution monitors (up to 2k x 2k pixel), frame buffers are usually implemented using slow memory chips with a very high density. Before Video RAM (VRAM) chips were introduced, only single ported dynamic RAM chips could be used to realize a frame buffer. Therefore the screen refresh consumed a significant fraction of the frame buffer access bandwidth, it became a bottleneck for the pixel update.

In general, modern applications are constructed of VRAM chips that provide an additional serial port [14,17]. Using this port for image refresh to the monitor, the random access port is nearly all of the time available for the pixel update (only a small amount of time is required for RAM refresh). So the rasterization process can be accelerated by simply using VRAM chips instead of DRAM chips.

Nevertheless, Computer Graphics users have an ever-increasing desire for higher performance, more functionality and better picture quality. This is constantly driven by rapid improvements in semiconductor technology and more sophisticated rasterization techniques [7]. Therefore the principal frame buffer functionality of storing pixels and refreshing the screen must be extended. Additional issues and their impacts on the overall performance have to be considered.

- Bringing parallelism to the frame buffer is an effective method to speed up image generation. On the pixel level the components of the pixel representation (e. g. color-, Z-value, window identifier) and on the primitive level several picture elements referring to the same graphics primitive (e. g. several spans of a triangle) can be computed and accessed in parallel. Additionally, the manipulation of geometric data may be performed by a parallel multiprocessor approach.

  In order to achieve this parallelism, various methods and techniques have been investigated [7]. As an example, the image memory can be organized into subsequent scan lines or square areas, and the memory chips can be arranged in an interleaved mode to reduce the influence of the memory cycle time. Furthermore, some frame buffer designs are based on custom designed VLSI chips which combine memory and processor capability on a single chip. The most prominent representatives are the Scan Line Access Memory (SLAM) [5], the Super Buffer [8] and Pixel-Planes [6]. They all are capable to cause a high speed-up for special rasterization algorithms. But problems of data distribution, flexibility, programmability, and mass production arise when systems with a very high pixel resolution have to be implemented, as it is required for modern workstations. Moreover, such graphics systems are mostly restricted to a fixed rasterization algorithm with the consequence that they cannot be adjusted to new and possibly more sophisticated techniques. As a result, the Distributed Frame Buffer is targeted as a state-of-the-art architecture [1,2,9,10]. It provides a high bandwidth between rendering processors and the frame buffer. This is achieved by enabling the pixel accesses of multiple rasterization units in parallel. Even in applications with a very high pixel resolution, an efficient design can be done by using off-the-shelf and hence cheap VRAM chips. A major part of this paper will focus on the distributed frame buffer concept as a basis for a high performance graphics system.

- In currently available workstations, a window management system is essential. Therefore, the appropriate frame buffer architecture should not only be optimized for fast primitive rendering within the whole screen area, but also within a screen portion - the window. This means that the throughput of processed graphics primitives should not decrease if the image generation is limited to the boundaries of a window. It should be kept in mind, that modern workstations even have to provide arbitrarily shaped windows [1,2].

  In summary the frame buffer architecture has to support efficiently fast window updates and movements, creating and deleting pop-up menus and handling arbitrarily shaped windows. In a multi-tasking environment these requirements have to be managed in conjunction with a fast context switch, because different window contents are usually processed by different tasks [11].

- The integration of video input sources into a graphics system by the use of a frame grabber hardware extension is a very reasonable feature for animation and image synthesis applications. The video input can then be manipulated by image processing algorithms and displayed like computer generated images. Nowadays, designers of electronic publishing systems investigate on the integration of moving images into electronic documents. Therefore a frame buffer has to provide a frame grabber mechanism. The windows containing video sequences should be controlled and manipulated in the same manner as a text or graphics window.

Keeping in mind, that the frame buffer must be tightly coupled to the rasterization processors, the strong influence of the frame buffer architecture on the hardware and software design of the overall system is becoming apparent. On the other hand, image generation, manipulation, and rasterization comprises many different tasks and complex computations. The stages of the image generation output pipeline can be qualified by a simple scheme as shown in figure 1. It is the basis for most of the graphics system architectures [2,3,9,16].

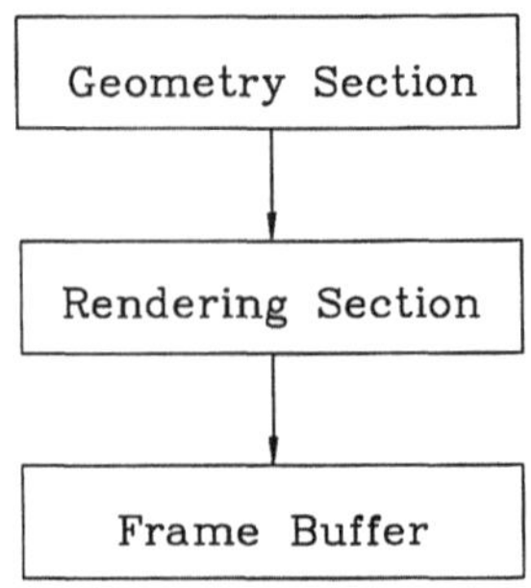

Figure 1: Image generation scheme

The geometry section comprises modelling and viewing transformations, the perspective projection and the clipping of the graphics primitives. The rendering section consists of the evaluation of an illumination model, of the rasterization process, shading and hidden surface removal. At the end of the rendering process the pixels are stored into the frame buffer and are displayed on the monitor.

A system architecture is needed where many processors perform all these tasks in order to achieve fast image generation and manipulation. With a parallel architecture the designer splits up the computational burden. Partitioning of the complete graphics process is state-of-the-art, and as a consequence there are a lot of proposals to design a graphics system with several processors working in parallel and executing different tasks.

One characteristic implementation is shown in [4] by developing hardware to speed up graphical algorithms of the geometry computation. It is the realization of a pipeline structure with all inherent disadvantages. It is almost impossible to balance the workload of the pipeline stages. Additionally the hardware pipeline is fixed for a given algorithm even if the requirements are changing.

Avoiding these problems Torborg, [15] presented a parallel architecture for the geometric section. The computational task is distributed to a configurable number of MIMD geometry processors. But even this solution presents a two stage pipeline consisting of the geometry and rendering section with possible difficulties of work-load balancing.

Therefore it is reasonable to apply Torborg's approach to the geometry and the rendering section. This leads to a well balanced system of several processors working in parallel, which takes advantage of the distributed frame buffer concept. It is realized in the 3DGRP - the 3D Geometry and Rendering Pipeline [13] which is also presented in this paper.

2. The 3DGRP Frame Buffer Architecture

The basic idea of the 3DGRP frame buffer architecture is to support a multiprocessor approach to speed up rendering algorithms in general. Although several rendering processors have to access the frame buffer in parallel to achieve the desired speed up a bottleneck must be avoided at the interface between the rendering processors and the memory chips of the frame buffer. This leads to the concept of the distributed frame buffer which is state of the art in modern workstations. For instance the raster subsystem of the Silicon Graphics' Superworkstation [1,2] is organized into five memory banks that are attached to one span processor respectively. The frame buffer in the AT&T Pixelmachine [9,10] is divided into an array of pixel nodes. Each pixel node contains a fraction of the whole memory tightly coupled to a MIMD processor.

For the 3DGRP frame buffer design three conditions for a resonable partition of the entire storage have to be taken into account:

- the required display resolution of 1280 x 1024 pixel
- the organization of modern VRAM-chips (256k x 4 bit)
- the number of necessary VRAM-chips, which should be minimal.

This leads to a distributed frame buffer consisting of five equally sized memory banks. Every bank contains 1280 : 5 = 256 columns of the whole screen. The arrangement of these columns is shown in figure 2.

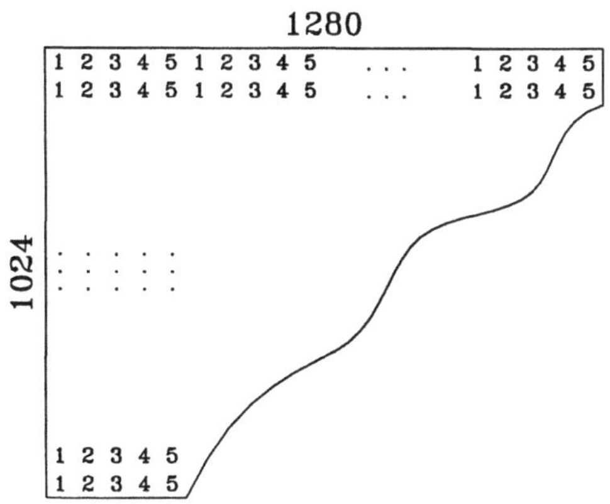

Figure 2: Frame buffer interleaving

Every memory bank can be accessed exclusively by a tightly coupled rendering processor, i.e. one processor need not to compete with others for memory access.

The global frame buffer architecture is represented by figure 3. The on-screen memory contains all the pixels visible on the screen. The resolution is 1280x1024 pixels. Each pixel consists of 24 bit color data twice (double buffer), 24 bit depth value (Z-buffer), 8 bit transparency, 8 bit window identifier, and up to 15 bit cursor overlay planes. The color representation can individually be chosen between true color and indexed color for every pixel. In case of true color the 24 bit color value will be splitted into the R-, G- and B-channel with 8 bit each. In the indexed color mode only 8 bits of the 24 bit color value are used as the input of the color look up table.

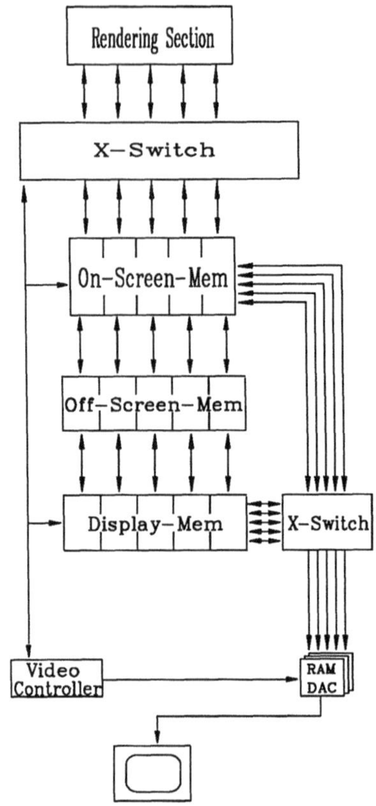

Figure 3: Frame buffer architecture

The off-screen-memory provides additional memory space for pixel data. It can be used by a window manager for the implementation of a backing store or as a font cache. The off-screen memory contains nearly two million pixels. The use of the display memory is described later on. All of these memories can be accessed by the rendering processors.

The display memory is used as temporary storage for all pixels within one or more window boundaries. It is needed for the hardware supported window copying mechanism described later on. All memory sections can be read and written by the rendering processors using the random-access ports of the memory chips, but the display memory cannot be seen by the user of the 3DGRP.

In order to realize BitBlT operations within the distributed frame buffer every processor needs to access all the frame buffer banks. Therefore a crossbar switch (X-switch) connects all memory banks with the rendering processors.

Every pixel has a 8 bit window identifier (window-ID) which is used to index a window look up table (WLUT). The window-ID controls write accesses of the rendering processors to the frame buffer as well as it presents basic information for the video controller. Figure 4 illustrates the video controller's evaluation mechanism of the window-ID.

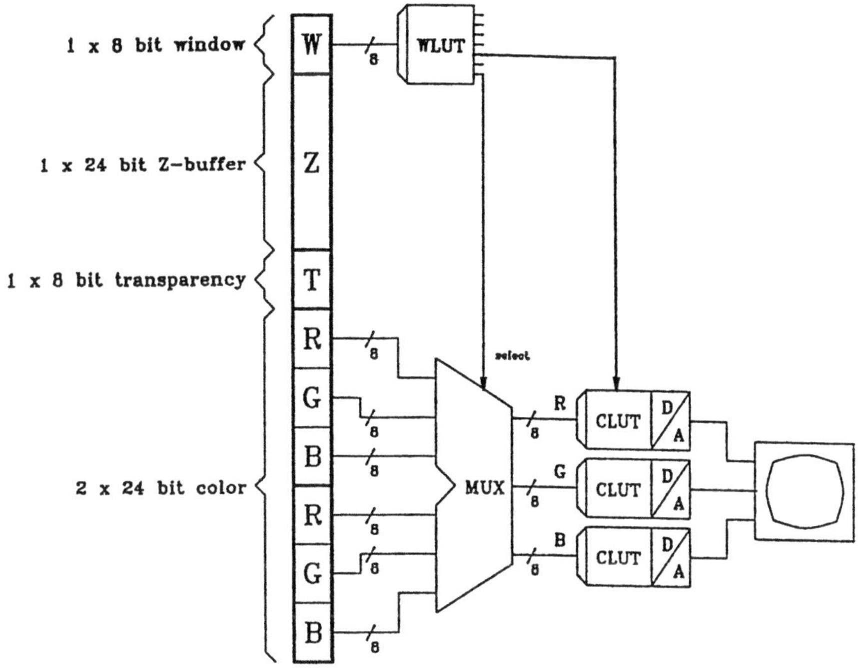

Figure 4: Window-ID evaluation

- Controlling the rendering processors' write accesses allows to handle arbitrarily shaped windows. Due to this for every write operation the rendering processors have to compare the window-ID of the new pixel with the stored one. In case of difference the write operation is cancelled. This mechanism is also used for off-screen memory access.
- The video controller utilizes the output of the WLUT to select for every window individually the source of pixel data which are to be displayed on the monitor Therefore double buffering is individually selected for every window.
- The window-ID manages to select one out of five colour look up tables (CLUT) for every window individually.
- The video controller can fill a window of any size and shape with a constant colour during one frame cycle by evaluating the WLUT output.
- Finally, advantage can be taken of the window-ID to support very efficiently the moving of multiple windows across the screen. Therefore the video controller is able to copy these windows without intervention of the processors. Before starting the copy function the window manager has to provide a shift vector, the identifiers of the windows to move and all identifiers of those windows which are allowed to cover the ones to move. While the copying is done by the video controller the rendering processors can already start to restore the formerly overlapped screen areas. The hardware copy function handles the visible part of the double buffer (color data), the window-IDs and optionally the cursor data.

The serial input/output ports of the VRAM chips of the on-screen and the display memory and the input ports of the RAMDACs are linked via the second crossbar switch. Normally the SAM ports of the on-screen memory are switched directly to the RAMDACs for a conventional screen refresh and to the SAM ports of the display memory to store all the pixel information of the present frame. This is a necessary precondition (figure 5a) before starting the hardware supported window copying described by the following example.

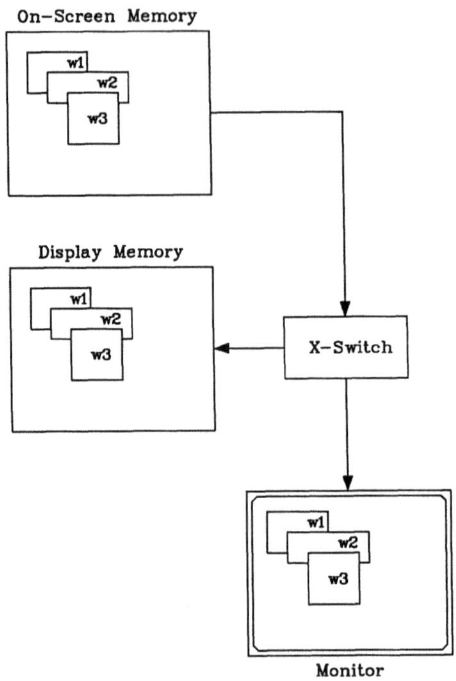

Figure 5a

A screen image including three windows W1, W2, W3 which is displayed on the monitor is stored in the on-screen and display memory. The joint windows W1, W2 are to be moved from the upper left corner (see figure 5b) in direction to the lower right corner. For this example the window W3 can occlude the windows W1 and W2 and the window W2 covers W1. The mechanism of the hardware supported window copying is illustrated by figure 5b to 5e.

The windows to be moved (W1, W2) are copied selectively into the display memory within one frame cycle. Simultaneously the contents of the on-screen memory is displayed on the monitor (figure 5b). Therefore the on-screen memory is read as in normal video refresh mode. The serial input stream to the display memory is controlled by two parameters: The window identifier related to every pixel determines (via the WLUT) if a pixel is to be overwritten in the display memory. The second parameter is a starting address of the display memory, which is calculated of a shift vector given by the window manager.

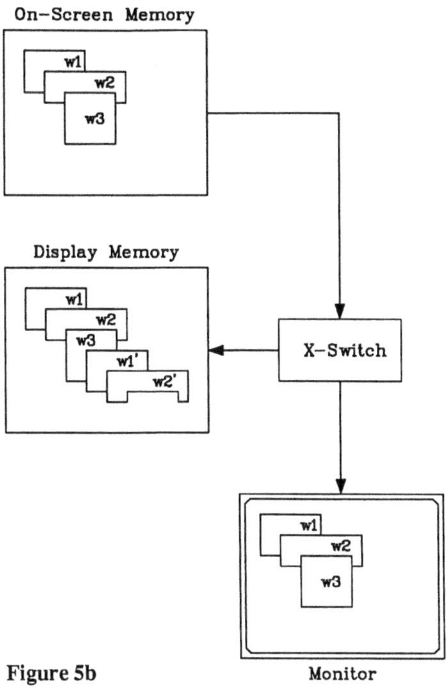

Figure 5b

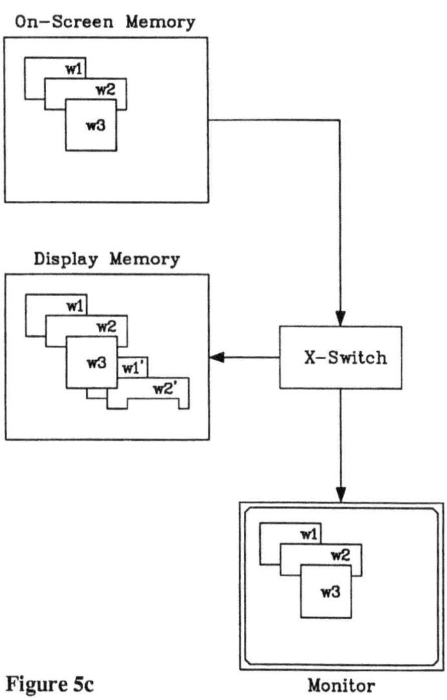

Figure 5c

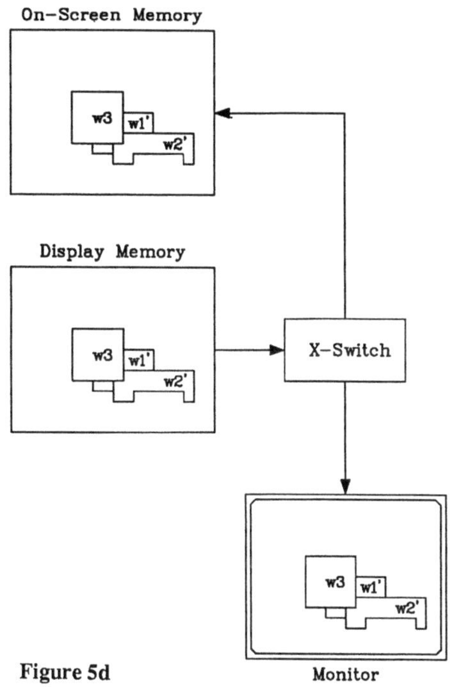

Figure 5d

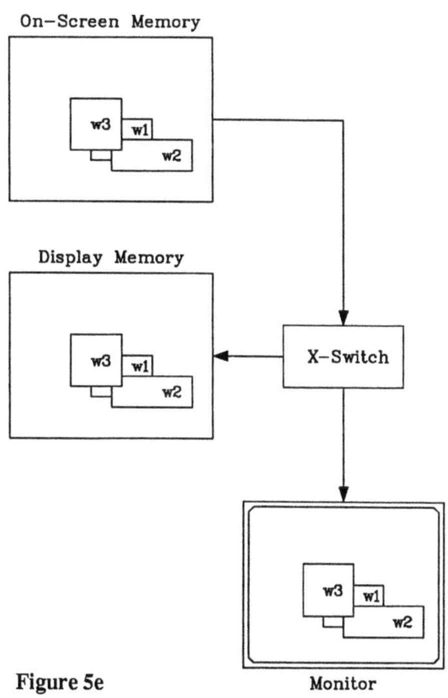

Figure 5e

In the following step all windows which may cover the ones to be moved (window W3 for this example) are copied into the display memory without regarding the shift vector(figure 5c). That means, the start addresses of the on-screen and display memory are the same. The selective copying is done as described in the previous step and again it lasts the time of one frame. The monitor input is still generated by the contents of the on-screen memory.

In the next frame, the video refresh is done by the display memory output. Simultaneously all pixels of the display memory are transferred via the serial links to the same location in the on-screen memory (figure 5d).

As a result of the copying procedure described above, there may be regions visible, which have to be regenerated by the rendering processors (e.g. background formerly covered by W1,W2). The rendering processors are free to perform this regeneration during all steps of the copying procedure which are illustrated by figure 5b to 5d. The final result is shown in figure 5e.

3. The 3DGRP Multiprocessor System

Realizing the disadvantages of existing architectures mentioned in the introduction, the goals for the 3DGRP are as follows:

- Highly parallel
  To achieve the required system performance.
- Homogeneous
  Only one type of processor shall be installed, in order to minimize the efforts for hardware and software development.
- Algorithm-independent
  Changing of algorithms must not require changes in the architecture.
- Application-independent
  Every application shall be performed with full system performance.
- Automatic load balancing
  No processor will enter a period of being idle, thus achieving the highest possible effectiveness.
- Easy to extend
  If a performance upgrade is needed, the hardware can be extended with minimal effort.

Figure 6 presents the global structure of the 3DGRP. There is a number of identical processor moduls GRi working in parallel and accessing the frame buffer. Each modul contains a processor, its memory and an interface to the geometry and rendering bus and to the frame buffer.

As in Torborg's approach, the objects are distributed to the processor moduls GRi via the geometry bus, which is implemented as multiprocessor bus as well as a broadcast bus. If a GR processor fetches a new geometric primitive to be processed, the geometry bus acts as a normal multiprocessor bus. Enabling the geometry bus in the broadcast mode it is a means to speed up the initialization phase and to shorten the overall transfer time. This is useful while updating context data within the memory of the GR modules.

Each processor performs the geometry processing for one object completely. The outcoming data are fed to the processors of the system via the rendering bus, which is similar to the geometry bus a broadcast bus as well as a normal multiprocessor bus. Then the processors render the objects for all those pixels being situated in the attached frame buffer bank. That means, the geometry processing of one object is done by only one processor. However, the rendering task of this object is done by all the processors but with reduced pixel amount.

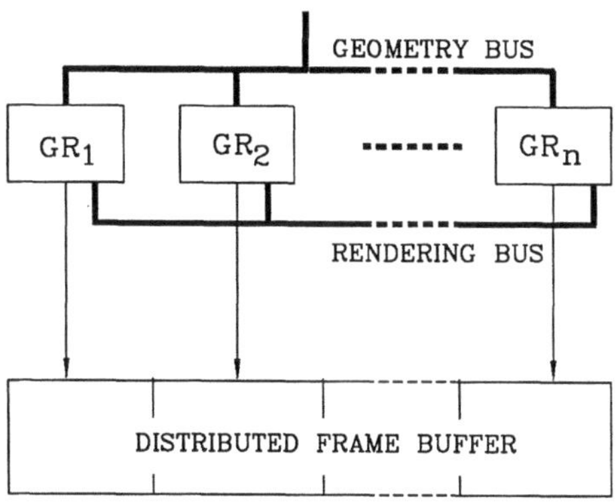

Figure 6: Global structure of the 3DGRP

The logical interface between the geometry and rendering calculations transfers pixels, vectors, triangles and trapeziums with edges parallel to the y-axis as rendering primitives to the GR modules. Due to their being preprocessed e. g. triangle data are calculated really fast. The start values for a span are computed in less than 500 ns and a pixel within a span is delivered every 1 us. Note that these values are valid for true color, z-buffered Gouraud shading, and all the calculations are done with full floating point precision, thus reducing computational inaccuracy to a minimum.

The effort of computation for processing the incoming data is dependent on the size and position of the geometric primitives. Small triangles or short vectors in parallel to the x- or y-axis require only a small number of rendering operations. The greater burden of computation relative to the resulting number of pixels of that primitive is in the geometry section. If there are very big triangles, much more processing power for rendering calculations is needed than for geometric calculations.

The peak performance of this architecture is limited by the number of processors involved and is delivered when nearly all processing power is exploited for rendering computations. The above mentioned data dependencies cause shifts of processing power requirements between the geometry and the rendering section. This architecture will overcome this difficulty within its area of performance.

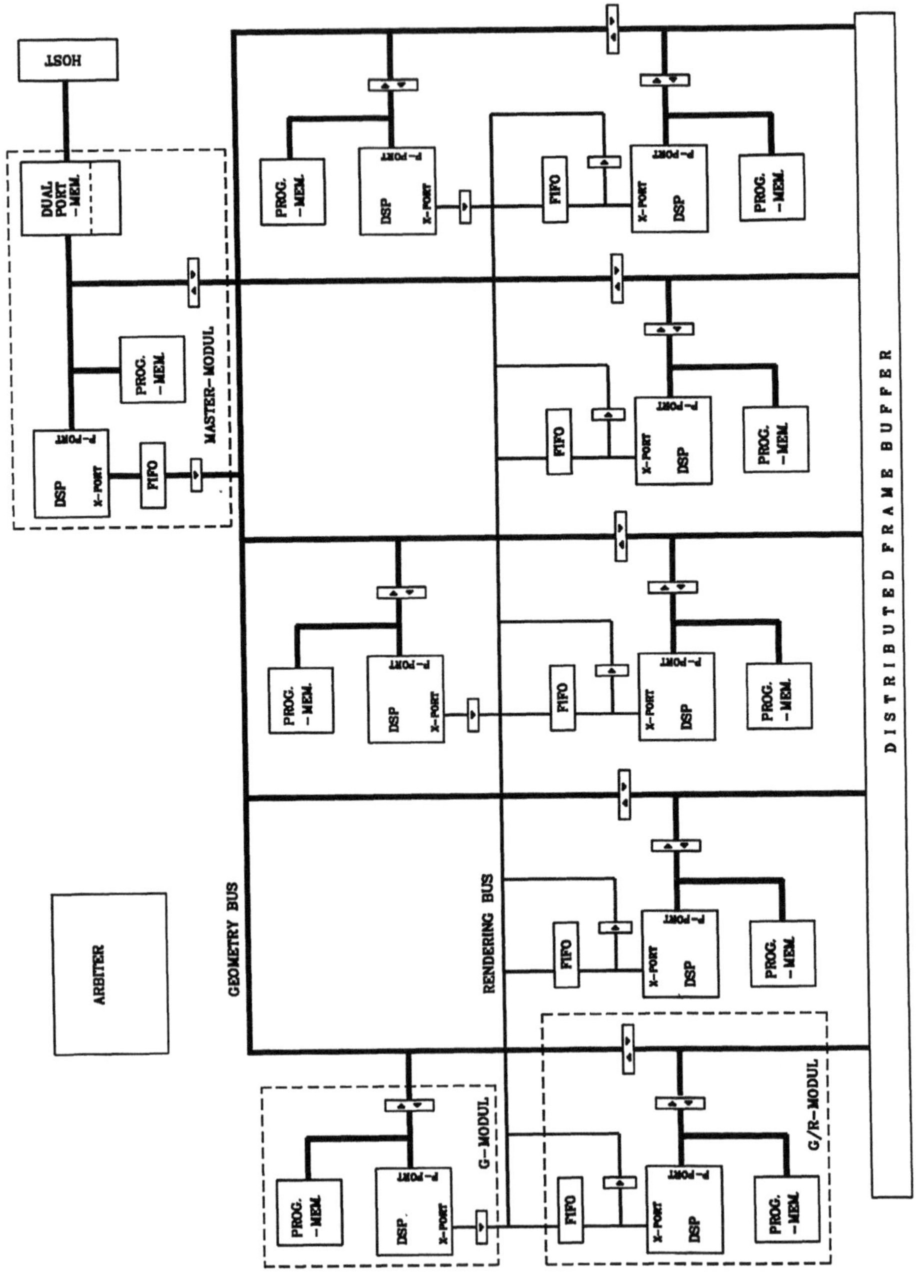

Figure 7: The 3DGRP architecture

Since any processor may run idle after rendering an object, if there is no rendering data in his input buffer, it will request new unprocessed objects and continue geometry calculations. In this way an automatic load-balancing is achieved across all the processors. When several GR moduls are doing geometric calculation, the overall rendering performance is reduced in favour of geometry processing power, but no computational power is going to be wasted by a GR processor starting out to run an idle state.

For this switching mechanism an algorithm is required, that changes quickly the context of computation. The internal structure of the processors (single cycle instructions, internal memory) provide a fast task switching within 2 us or less thus supporting sustained processor performance and avoiding the advances of automatic load balancing to be eaten up by switching time.

In figure 7 a realization example of 3DRP is given, utilizing five GR-modules. As processor the digital signal processor (DSP) Texas TMS 320C30, the fastest microprocessor available at present, has been chosen. This processor provides two memory ports, so it can be connected easily to the geometry and rendering bus.

For graphics applications where the computation burden within the geometry section is much higher than within the rendering section the basic structure shown in figure 6 can be augmented with several geometry moduls (G modul). In figure 7 there are three G moduls presented. The processor of a G modul performs only the geometry tasks of the graphics application. Therefore the G modul is a copy of the GR modul, but a FIFO and a frame buffer interface are not needed.

The master DSP within the master modul (fig. 7) communicates with the host via a dual-ported memory. The host transfers the graphics data of the application into the dual-ported memory. The master DSP interprets and preprocesses the data and moves them via the extension port into the master FIFO. All processors (of the G and GR moduls) can read the geometric data from the master FIFO. The arbiter in the background is scheduling the geometry bus, so that only one processor at a time can access the geometry bus.

The performance of this realization of a 3DGRP architecture is estimated as nearly 400,000 Gouraud-shaded 10 pixel vectors per second and nearly 50,000 Gouraud-shaded 100 pixel triangles per second.

References

1. Akely, K., Jermoluk, T.: High Performance Polygon Rendering. Proceedings of SIGGRAPH, 22(4): pp. 239-246, August 1988
2. Akely, K.: The Silicon Graphics 4D/240GTX Superworkstation. IEEE Computer Graphics & Applications, 7: pp. 71-83, July 1989
3. Apgar, B., Bersack, B. and Mammen, A.: A Display System for the Stellar Graphics Supercomputer Model GS1000. Proceedings of SIGGRAPH, 22(4): pp. 255-268, August 1988
4. Clark, J.: The Geometry Engine: A VLSI Geometry System for Graphics. Computer Graphics, 16(3): pp. 127-133, July 1982
5. Demetrescu, S.: High Speed Image Rasterization Using Scan Line Access Memories. Proc. 1985 Chapel Hill Conference on VLSI, pp. 221-243, Computer Science Press, 1985
6. Fuchs, H. and Poulton, J.: Pixel Planes: A VLSI-Oriented Design for a Raster Graphics Engine. VLSI Design, 2(3): pp. 20-28, 3rd. Quater 1981
7. Gharachorloo, N., Gupta, S., Sproull, R.F. and Sutherland, I.E.: A Characterization of Ten Rasterization Techniques. Proceedings of SIGGRAPH, 23(3): pp. 355-368, July 1989
8. Gharachorloo, N. and Pottle, C.: SUPER BUFFER: A Systolic VLSI Graphics Engine for Real Time Raster Image Generation. Proc. 1985 Chapel Hill Conference on VLSI, pp. 285-305, Computer Science Press, 1985
9. Potmesil, M. and Hoffert, E.M.:The Pixel Machine: A Parallel Image Computer. Proceedings of SIGGRAPH, 23(3): pp. 69-78, July 1989
10. Potmesil, M., McMillan, L., Hoffert, E.M. et al.: A Parallel Image Computer with a Distributed Frame Buffer: System Architecture and Programming. Proceedings of EUROGRAPHIGS '89, Hamburg, FRG: pp. 197-208, September 1989
11. Rhoden, D., Wilcox, C.:Hardware Acceleration for Window Systems. Proceedings of SIGGRAPH, 23(3): pp. 61-67, July 1989
12. Rogers, D.F.: Procedural Elements for Computer Graphics. MacGraw-Hill, 1985
13. Selzer, H., Haaker, T., Joseph, H.: 3DGRP - A High Performance Graphics System. Proceedings of The 1989 IFIP WG5.10 International Working Conference on Workstations for Experiments, Lowell, USA, July 1989
14. Sproull, R.F.: Frame Buffer Display Architectures. Annual Review of Computer Sience, 1: pp. 19-46, Annual Reviews Inc., 1986
15. Torborg, J.: A Parallel Processor Architecture for Graphics Arithmetic Operations. Computer Graphics, 21(4): pp. 197-204, 1987
16. Tunick, D.: Powerful display system revs up image and graphics processing. Electronic Design, July 23, 1987
17. Whitton, M.C.: Memory Design for Raster Graphics Displays. Computer Graphics & Applications, 4(3): pp. 48-65, March 84

List of Participants[1]

Mark Agate
School of Engineering
University of Sussex
Brighton, BN1 9QT
United Kingdom

Anthony Barkans
Graphics Technology Division
Hewlett Packard
Ft. Collins, Colorado
USA
email: barkans@hpfcre

France Glémot
CAPTION SARL
23 Rue du Bignon
35 135 Chantepie
France

Richard Grimsdale
School of Engineering
University of Sussex
Brighton, BN1 9QT
United Kingdom
email: dickg@syma.sussex.ac.uk

Rajiv Gupta
256-80 Caltech
CA 91125 Pasadena
USA
email: gupta@csvax.caltech.edu

T. Haaker
FB Graphisch-Interaktive Systeme
TH Darmstadt
Wilhelminenstr. 7
6100 Darmstadt
FRG

Dietmar Jackél
Fachbereich 20, Fr 3-3
TU Berlin
1000 Berlin
FRG

Shi Jiaoying
Dept. of Computer Science &
Engineering
Zhejiang University
China

Arie Kaufman
Dept. of Computer Science
State University of New York
Stony Brook, NY 11794-4400
USA
email: ari@sbcs.sunysb.edu

Ulrich Kursawe
Development Dept.
Kontron Elektronik
Breslauer Str. 2
8057 Eching
FRG

[1]For Non-commercial use only

Paul Lister
School of Engineering
University of Sussex
Brighton, BN1 9QT
United Kingdom
email: paull@syma.sussex.ac.uk

Claudia Romanova
Wilhelm-Schickard-Institut
Graphisch-Interaktive Systeme
Universität Tübingen
Auf der Morgenstelle 10, C9
7400 Tübingen
FRG
email: claudia@gris.informatik.uni-tuebingen.de

Bengt-Olaf Schneider
Wilhelm-Schickard-Institut
Graphisch-Interaktive Systeme
Universität Tübingen
Auf der Morgenstelle 10, C9
7400 Tübingen
FRG
email: bengt@gris.informatik.uni-tuebingen.de

Wolfgang Straßer
Wilhelm-Schickard-Institut
Graphisch-Interaktive Systeme
Universität Tübingen
Auf der Morgenstelle 10, C9
7400 Tübingen
FRG

John Vaudin
Dept. of Computer Science
Warwick University
Coventry, CV4 7AL
United Kingdom

Celal Yilmaz
J. Jittastr. 40
3052 PR Rotterdam
The Netherlands

Masaharu Yoshida
NTT
Human Interface Labs.
1-2356 Yokosuka-chi
Kanagawa
Japan
hal%scandium.ntt.jp

EurographicSeminars

Tutorials and Perspectives in Computer Graphics

Eurographics Tutorials '83. Edited by P. J. W. ten Hagen.
XI, 425 pages, 164 figs., 1984

User Interface Management Systems. Edited by G. E. Pfaff.
XII, 224 pages, 65 figs., 1985

Methodology of Window Management. Edited by F. R. A. Hopgood, D. A. Duce, E. V. C. Fielding, K. Robinson, A. S. Williams.
XV, 250 pages, 41 figs., 1985

Data Structures for Raster Graphics. Edited by L. R. A. Kessener, F. J. Peters, M. L. P. van Lierop. VII, 201 pages, 80 figs., 1986

Advances in Computer Graphics I. Edited by G. Enderle, M. Grave, F. Lillehagen.
XII, 512 pages, 168 figs., 1986

Advances in Computer Graphics II. Edited by F. R. A. Hopgood, R. J. Hubbold, D. A. Duce. X, 186 pages, 96 figs., 1986

Advances in Computer Graphics Hardware I. Edited by W. Straßer.
X, 147 pages, 76 figs., 1987

GKS Theory and Practice. Edited by P. R. Bono, I. Herman.
X, 316 pages, 92 figs., 1987

Intelligent CAD Systems I. Theoretical and Methodological Aspects.
Edited by P. J. W. ten Hagen, T. Tomiyama. XIV, 360 pages, 119 figs., 1987

Advances in Computer Graphics III. Edited by M. M. de Ruiter.
IX, 323 pages, 247 figs., 1988

Advances in Computer Graphics Hardware II. Edited by A. A. M. Kuijk, W. Straßer.
VIII, 258 pages, 99 figs., 1988

CGM in the Real World. Edited by A. M. Mumford, M. W. Skall.
VIII, 288 pages, 23 figs., 1988

Intelligent CAD Systems II. Implementational Issues. Edited by V. Akman, P. J. W. ten Hagen, P. J. Veerkamp. X, 324 pages, 114 figs., 1989

Advances in Computer Graphics IV. Edited by W. T. Hewitt, M. Grave, M. Roch.
XVI, 248 pages, 138 figs., 1991

Advances in Computer Graphics V. Edited by W. Purgathofer, J. Schönhut.
VIII, 223 pages, 101 figs., 1989

User Interface Management and Design. Edited by D. A. Duce, M. R. Gomes, F. R. A. Hopgood, J. R. Lee. VIII, 324 pages, 117 figs., 1991

Advances in Computer Graphics Hardware III. Edited by A. A. M. Kuijk. VIII, 214 pages, 88 figs., 1991

Advances in Object-Oriented Graphics I. Edited by E. H. Blake, P. Wisskirchen. X, 218 pages, 74 figs., 1991

Advances in Computer Graphics Hardware IV. Edited by R. L. Grimsdale, W. Straßer. VIII, 276 pages, 124 figs., 1991

In preparation:

Advances in Computer Graphics VI. Edited by G. Garcia, I. Herman. Approx. 465 pages, 1991

Intelligent CAD Systems III. Practical Experience and Evaluation. Edited by P. J. W. ten Hagen, P. J. Veerkamp. Approx. 280 pages, 1991

MIX
Papier aus verantwortungsvollen Quellen
Paper from responsible sources
FSC® C105338

If you have any concerns about our products,
you can contact us on
ProductSafety@springernature.com

In case Publisher is established outside the EU,
the EU authorized representative is:
Springer Nature Customer Service Center GmbH
Europaplatz 3, 69115 Heidelberg, Germany

Printed by Libri Plureos GmbH
in Hamburg, Germany